# Foundation
# Electrical
# Engineering

# *Foundation Electrical Engineering*

J. P. Mc TAVISH

LIVERPOOL JOHN MOORES UNIVERSITY

PRENTICE HALL

*London   New York   Toronto   Sydney   Tokyo   Singapore*
*Madrid   Mexico City   Munich*

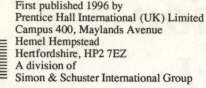

First published 1996 by
Prentice Hall International (UK) Limited
Campus 400, Maylands Avenue
Hemel Hempstead
Hertfordshire, HP2 7EZ
A division of
Simon & Schuster International Group

Typeset in 10/12 pt Times
by Mathematical Composition Setters Ltd, Salisbury

Printed and bound in Great Britain by
T. J. Press (Padstow) Ltd

Library of Congress Cataloging-in-Publication Data

Mc Tavish, J. P.
    Foundation electrical engineering / J.P. Mc Tavish.
       p.  cm.
    Includes index.
    ISBN 0-13-309931-8 (alk. paper)
    1. Electric engineering.   I. Title
  TK146.M43   1995
  621.3–dc20                    95-20081
                                      CIP

British Library Cataloguing in Publication Data

A catalogue record for this book is available from
the British Library

ISBN 0-13-309931-8

1  2  3  4  5    00  99  98  97  96

# Contents

# *Preface*

The Foundation Course at Liverpool John Moores University (LJMU) is designed to enable people without the traditional entrance requirements in mathematics and a science subject, but have reached the same educational level in other 'non-scientific' subjects, to enter the first year of a wide range of engineering or applied science degrees.

The course consists of two semesters in which Mathematics, Mechanical Engineering Science, Electrical Engineering Science, and Engineering Appreciation or Modern Physics are studied

It is the purpose of this text to provide an introduction to electrical engineering science which will enable the reader to approach the first year of a degree in Electrical Engineering with confidence. For this it is necessary to be able to perform only some surprisingly simple calculations involving the most rudimentary mathematics, namely addition, subtraction, multiplication and division. Although most of the mathematics is straightforward, this does appear to be a stumbling point with many students at this level. The author believes one of the reasons for this is that the subject is often presented in terms of rules with only the briefest of explanations; for example, to find the current flowing through a circuit just divide one of the numbers by another. By presenting some explanation or background to the calculation it becomes more palatable as to why these certain numbers are manipulated in a particular fashion – the calculation becomes less mysterious and more reasonable. Thus, the term *reasonable* is used quite often throughout the test. This is deliberate, since once it is accepted, for example, that when an electric current passes through a conductor there will be a heating effect, applying the calculation to determine the rate at which heat is generated

$$\text{power dissipated} = \text{current} \times \text{potential difference}$$

becomes a matter of common sense and is part of the process of performing the calculation. The other important part of that process is actually being able to perform that calculation. This is the reason for the appearance of the Worked Examples throughout the text. The worked examples are therefore a very important part of the text. They provide practical applications of the theory discussed previously. Some of the Worked Examples are just straightforward applications of particular formulae, while others are more involved and provide realistic calculations that will be met in the first year of a degree course.

Before discussing electricity it is first useful to discuss units. All quantities we make use of have two parts: a number part and a unit part. In this text we make use of a particular set of measuring units referred to as the 'International System of Units' (SI). This is an internationally agreed measurement system which is widely used throughout the scientific and engineering world. We also take the opportunity to introduce standard and scientific notations, which are merely convenient and widely used ways of expressing a number.

We begin our study of electric circuits by introducing electric charge in terms of something that explains the experimental results of rubbing materials with a piece of cloth. After this, the notion of electric charge in motion is introduced, which will be termed an *electric current*, and the connection with the familiar term *electricity* is established. We then proceed to discuss electric circuits proper. For this we need to introduce resistance and its effect on a circuit. In particular, we are interested in how resistance limits the current through a circuit. We will find that the link between resistance and current is the *electromotive force* of the battery. Once this link has been established we are able to discuss a wide range of electric circuits. We then go on to present some methods, in particular Kirchhoff's laws, the principle of superposition and Thevenin's theorem, which enable us to consider a circuit of arbitrary complexity, though we discuss only simple circuits that bring out the essence of the methods. We then discuss a subject of great practical importance, namely the magnetic effect of an electric current. This introduces the related subject of inductance and establishes a link between the subjects of electricity and magnetism which is referred to as electromagnetism. After discussing capacitors we then go on to a discussion of alternating current.

A subject closely related to electricity is the subject of *electronics*. Where electricity ends and electronics begins differs according to the text being read. We will take electronics to be the study of *active* electrical components. For us this will mean that electronics begins with our discussion of the transistor. Before discussing the transistor we first need to discuss semiconductors and the diode.

The text has been written with the assumption that the reader has no prior knowledge of electricity and only the barest knowledge of mathematics. This will enable the student to get through the first few chapters. To proceed further, it is assumed that the reader is taking suitable mathematics courses. In particular, some knowledge of differentiation and integration, particularly of the sine and cosine functions, are needed. Even then, the results are given in the text and all that is really required is that the reader appreciate the meaning of what these mathematical operations entail. A mathematical appendix is supplied which presents material which is of direct use. This should be consulted as required. Finally, material in the text marked with a circle in the margin (page 21, for example) represents advanced concepts or material which is not important for the development of the course, and can be omitted if so desired.

# *Acknowledgements*

The author would like to thank Mr A.F. Abbott for permission to make use of the following figures from *Ordinary Level Physics* by A.F. Abbott, Heinemann Educational Books, London (third edition, 1977): Figure 36.4 (page 429), Figure 36.5 (page 429), Figure 36.6 (page 430), Figure 36.16 (page 434), Figure 38.3 (page 449), Figure 38.4 (page 449), Figure 38.6 (page 451), Figure 38.7 (page 451), Figure 43.1 (page 498), Figure 43.5 (page 501), Figure 43.15 (page 508), and Figure 43.19 (page 511).

Thanks must also go to the following colleagues and friends who have helped in various ways: Professor R. Morgan (EEE, LJMU), Mr F. Smith (ETM, LJMU), Mr D. Armes (Southport College) and Mr K. Wong (Mathematics Department, University of British Columbia).

The author also acknowledges the support and interest in the text from the following: Dr J.T. Atkinson (EEE, LJMU), Mr P. Blackmore (EEE, LJMU), Mr S. Caulder (EEE, LJMU), Mr C. Clarke (EEE, LJMU), Mr M. Earner (EEE, LJMU), Mr R. Evans (EEE, LJMU), Mr M. Golding, Mr M. Harrington (EEE, LJMU), Mr L. Hughes (EEE, LJMU), Mr H. Kant (EEE, LJMU), Mr W. Nolan (EEE, LJMU), Mr S. Walker (CSD, LJMU) and Mr R. Wilson (EEE, LJMU), Mr B. Williams (LJMU) and Mr G. Wong (ETM, LJMU).

The help of Mr C. Glennie and Ms L. Wilson of Prentice Hall is also acknowledged.

Special thanks must also go to Mr M. Ahmed (NOWREC, LJMU), Mr D. Ellis (NOWREC, LJMU), Mr S. Harlow (Senior Fluid Transfer Technician, Shell, Sechelt, British Columbia) and Mr J. Windle (EEE, LJMU).

# CHAPTER ONE

# *An introduction to units and notation*

In this chapter we introduce first the SI system of units and then discuss standard and scientific notations. The International System of Units (SI) is a particular internationally agreed measurement system which is used widely throughout the scientific and engineering world. Throughout science in general and engineering in particular we often find we have to deal with both very large and very small numbers. The standard and scientific notations are merely convenient and widely used ways of expressing a number.

## 1.1 The International System of Units (SI)

Until about the year 1800 workers in various countries used different systems of units to measure the same quantity. For example, while scientists and engineers in England would measure lengths in inches their European counterparts would use centimetres. When comparing work it would then be necessary to convert each other's results to the particular system of units being used. This was obviously a very inconvenient and cumbersome arrangement. Due to the efforts of various international committees this situation gradually changed and in 1960 it was agreed that general use should be made of a system of measurement called the International System of Units (SI).

In this section we present a brief review of the SI system of measurement and units: The SI system includes seven *base* units. These base units are given in Table 1.1. Two supplementary units are used to measure angles, namely the radian (rad) for ordinary angles and the steradian (sr) for solid angles. In our work the first five are used most and we will restrict our attention to these.

Units for all other quantities are *derived* from these base units, i.e. they are built up from different combinations of the base units. In the following, the reader should not worry unduly at this stage if the scientific arguments or equations are not understood precisely. The important point to appreciate is that we can analyze the relationship between quantities in terms of their constituent base units. In mechanics,

1

*Table 1.1*

| Base unit | Name | Symbol |
|-----------|------|--------|
| length | metre | m |
| mass | kilogram | kg |
| time | second | s |
| electric current | ampere | A |
| temperature | kelvin | K |
| amount of substance | mole | mol |
| luminosity | candela | cd |

for example, we make use of only the first three units, namely the metre, kilogram and the second. Some of the more common derived units are given in Table 1.2.

From Table 1.2 we have, for example, that the SI unit of area is the square metre. This follows since to determine the area of a figure we multiply one length (measured in m) by another length. For example, the area of a rectangle of side 0.5 m and 2.0 m is given as

$$\text{area} = (0.5 \text{ m}) \times (2.0 \text{ m})$$
$$= (0.5 \times 2.0) \text{ m} \times \text{m}$$
$$= 1.0 \text{ m}^2$$

Note that in the calculation the number 0.5 multiplies the number 2.0 to give 1.0, while the unit metre multiplies the unit metre to give the unit square metre. This can be written as m × m but it is more usual to write as m², where the superscript 2 means 'multiply the expression on the left-hand side by itself'. The meaning is the same when used with numbers; for example, $5^2$ means $5 \times 5 = 25$. Similarly, we have that volume is measured in cubic metres, i.e. m × m × m or m³.

From Table 1.2 we have that speed is a measure of the change of position with time which has the SI of metre per second, m/s or m s⁻¹. In this example, note that we have made use of the notation

$$\frac{1}{s} \equiv s^{-1}$$

*Table 1.2*

| Derived unit | Name | Symbol |
|--------------|------|--------|
| area | square metre | m × m or m² |
| volume | cubic metre | m × m × m or m³ |
| speed | metre per second | m/s or m s⁻¹ |
| acceleration | metre per square second | m/s² or m s⁻² |
| momentum | kilogram metre per second | kg m/s or kg m s⁻¹ |

so that we can write the unit of speed in the form

$$\frac{m}{s} \equiv m\,s^{-1}$$

This is important notation to get used to since it is used extensively in both the scientific and engineering worlds.

Some derived units which have special names and which will be important for future discussion are given in Table 1.3.

When we are dealing with an equation that gives the relationship between different quantities, we must necessarily have that the units on both sides of the equation are the same. This is demonstrated in Worked Examples 1.1 and 1.2.

■■■■ **WORKED EXAMPLE 1.1**

If a body starts off from rest under a constant acceleration $a$, its final speed $v$ after travelling a distance $s$ can be shown to be given by the result

$$v^2 = 2as \tag{1.1}$$

Use this result to determine the SI unit of acceleration.
– From Table 1.1 distance is one of the SI base units and has the unit of m, while Table 1.2 gives the units of speed in terms of the base units length and time as

$$m\,s^{-1}$$

From the equation relating speed, acceleration and distance we have

units of $v^2$ = (units of acceleration) × (units of distance)

so that

$$\text{units of acceleration} = \frac{\text{units of } v^2}{\text{units of distance}}$$

Using as the unit of speed the metre per second, i.e. $m\,s^{-1}$, we have the units of $v^2$ as

$$(m\,s^{-1})^2 = (m\,s^{-1} \times (m\,s^{-1})$$
$$= (m \times m) \times (s^{-1} \times s^{-1})$$
$$= m^2 s^{-2}$$

so that we find

$$\text{units of acceleration} = \frac{m^2\,s^{-2}}{m} = \frac{m \times \cancel{m} \times s^{-2}}{\cancel{m}} = m\,s^{-2}$$

There are two important points to note:

- in equations, units are treated in the same manner as numbers; and
- in Equation (1.1) the factor of '2' is a pure number and is taken to have no units.

■■■■ **WORKED EXAMPLE 1.2**

Given that the relationship between the force, *F*, acting on a body of mass *m* and its acceleration *a* is

$$F = ma$$

determine the relationship between the units of force and those of mass and acceleration.
− From Table 1.1, mass is one of the SI base units and has the unit of kg. The units of acceleration in terms of the base units length and time was determined in Worked Example 1.1 as

$$m\,s^{-2}$$

From the defining equation it follows that the units of force are the same as the units of mass multiplied by the units of acceleration, i.e.

(units of mass) × (units of acceleration) = $kg\,m\,s^{-2}$

as given in Table 1.3.

An important point to note is that we have used the symbol *m* to be both the mass of an object and the base unit metre. This problem occurs often in science and engineering and it is important that the correct meaning should be attributed to a symbol. This meaning should be obvious from the context in which the symbol is being used. This problem is helped by adopting the convention of using italics to denote variables and roman type to denote units and labels. For example, we denote mass and electric charge by the symbols *m* and *Q* respectively, while we denote the metre and coulomb by the symbols m and C. Similarly, when we discuss atomic structure, we will denote the first three atomic shells by the labels K, L and M, respectively.

*Table 1.3*

| Derived unit | Name | Symbol | Derivation |
|---|---|---|---|
| force | newton | N | $kg\,m\,s^{-2}$ |
| energy | joule | J | $kg\,m^2\,s^{-2}$ |
| power | watt | W | $kg\,m^2\,s^{-3}$ |
| electric charge | coulomb | C | $A\,s$ |
| electric potential | volt | V | $kg\,m^2\,s^{-3}\,A^{-1}$ |
| resistance | ohm | Ω | $kg\,m^2\,s^{-3}\,A^{-2}$ |
| magnetic flux | weber | Wb | $kg\,m^2\,s^{-2}\,A^{-1}$ |
| magnetic field strength | tesla | T | $kg\,s^{-2}\,A^{-1}$ |
| inductance | henry | H | $kg\,m^2\,s^{-2}\,A^{-2}$ |
| capacitance | farad | F | $kg^{-1}\,m^{-2}\,s^4\,A^2$ |
| frequency | hertz | Hz | $s^{-1}$ |

## 1.2 Standard notation

Many quantities that we deal with have either very large or very small values. For example, the radius of the Earth is approximately 6500000 m, while the radius of a red blood cell is approximately 0.000004 m. This method of recording the data is cumbersome both to read and to write. It is slightly improved by the practice of grouping the individual digits into groups of three, i.e. 6500000 m and 0.000004 m, respectively, though the following method, referred to as standard notation, is actively used.

A very convenient method of dealing with this problem is to make use of powers of 10. For numbers larger than or equal to one we use the following notation:

$$10^0 = 1$$
$$10^1 = 10$$
$$10^2 = 10 \times 10 = 100$$
$$10^3 = 10 \times 10 \times 10 = 1000$$
$$10^4 = 10 \times 10 \times 10 \times 10 = 10\,000$$

and so on, in an obvious fashion. The number of zeros corresponds to the power to which 10 is raised, and is called the exponent. For example, the radius of the Earth given as 6500000 m is more conveniently written as $6.5 \times 10^6$ m. This is shown in greater detail in Figure 1.1.

For numbers less than one we have to make use of *negative* exponents, as follows:

$$10^{-1} = \frac{1}{10} = 0.1$$

$$10^{-2} = \frac{1}{10 \times 10} = 0.01$$

$$10^{-3} = \frac{1}{10 \times 10 \times 10} = 0.001$$

$$10^{-4} = \frac{1}{10 \times 10 \times 10 \times 10} = 0.000\,1$$

$$10^{-5} = \frac{1}{10 \times 10 \times 10 \times 10 \times 10} = 0.000\,01$$

In these cases the number of places the decimal point is to the left of the digit 1 is equal to the value of the negative exponent. Note that

$$10^{-m} = \frac{1}{10^m} \tag{1.2}$$

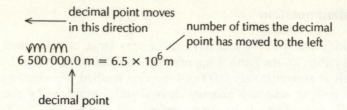

Figure 1.1

Making use of negative exponents we can write the radius of the blood cell, given as $0.000\,004$ m, as $4 \times 10^{-6}$ m. This is shown in greater detail in Figure 1.2.

In general we can write any number in the following form:

$$a \times 10^b$$

The number $a$ is called the mantissa or multiplying factor and is between 1 and 10, while $b$ is called the exponent and can be any positive or negative integer. This way of writing down a number is referred to as

*standard notation*

When numbers expressed in standard notation are being multiplied the following rule is very useful:

$$10^m \times 10^n = 10^{m+n} \tag{1.3}$$

This rule applies for all $m$ and $n$, in particular even if $m$ or $n$ are negative.

When numbers expressed in standard notation are being divided the following rule is very useful:

$$\frac{10^n}{10^m} = 10^n \times 10^{-m} = 10^{n-m} \tag{1.4}$$

where we have made use of the result (1.2).

■■■ **WORKED EXAMPLE 1.3**

Express the following numbers in standard notation:

(a) $0.000\,000\,000\,052\,9$ m,    (b) $299\,792\,458$ m s$^{-1}$.

– Standard notation suggests we write each of the numbers in the form

$$a \times 10^b$$

where $1 < a < 10$ and there are no restrictions on $b$. Using this rule we have

(a) $5.29 \times 10^{-11}$ m,    (b) $2.997\,924\,58 \times 10^8$ m s$^{-1}$.

The Worked Example 1.4 considers some simple mathematical operations in which numbers are expressed in standard notation.

━━━ **WORKED EXAMPLE 1.4**

Express the following numbers in standard notation:

(a) $(2.8 \times 10^3) \times (6.0 \times 10^4)$,    (b) $1.65 \times 10^4 + 1.8 \times 10^3$.

— Standard notation suggests we write each of the numbers in the form

$a \times 10^b$

where $1 < a < 10$ and there are no restrictions on $b$. Using this rule
(a) We can write

$(2.8 \times 10^3) \times (6.0 \times 10^4) = (2.8 \times 6.0) \times (10^3 \times 10^4)$

where we have first multiplied the two multiplying factors together, i.e. $2.8 \times 6.0$, and then the two powers of ten together, i.e. $10^3 \times 10^4$. Making use of the rule (1.3) we find the answer is given as

$16.8 \times 10^7$

This is not quite in standard notation since according to the definition, the multiplying factor must be between 1.0 and 10.0. Making use of the fact that we can write

$16.8 = 1.68 \times 10$

we can make use of the rule (1.3) again to write the final answer as

$1.68 \times 10^8$

(b) To add two numbers in standard notation together when the exponents are equal is straightforward; we merely add the two multiplying factors together and then multiply by the respective power of ten. When the exponents are unequal, as in this example, we simply convert one of the numbers to have the same multiplying power of ten as the other. For example, we can write

$1.65 \times 10^4 = 1.65 \times 10^1 \times 10^3 = 16.5 \times 10^3$

so that the sum can now be written as

$16.5 \times 10^3 + 1.8 \times 10^3$

giving the answer as

$(16.5 + 1.8) \times 10^3 = 18.3 \times 10^3$

and where, as in part (a), the next step is to convert the answer into standard notation by ensuring the multiplying factor is in the required range.

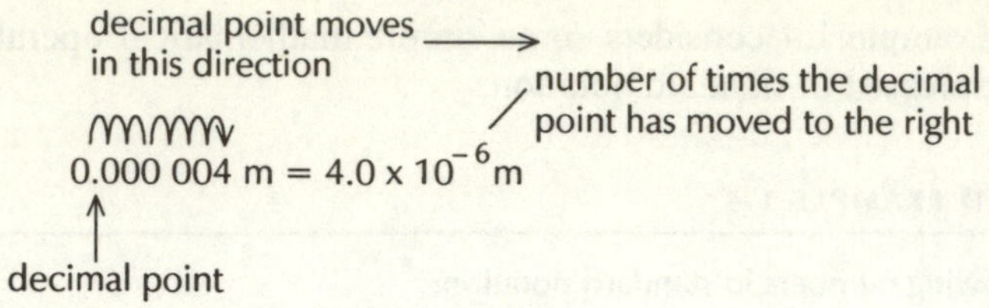

decimal point moves
in this direction ⟶

number of times the decimal
point has moved to the right

$0.000\ 004\ m = 4.0 \times 10^{-6}\,m$

decimal point

*Figure 1.2*

▄▄▄▄ **WORKED EXAMPLE 1.5**

Consider the following calculation:

$$\frac{0.000\ 43 \times 500 \times 0.129}{0.022\ 3}$$

— At first glance it is not obvious whether we expect the answer to be small or large. Making use of standard notation we can rewrite the above as

$$\frac{(4.3 \times 10^{-4}) \times (5.0 \times 10^{2}) \times (1.29 \times 10^{-1})}{2.23 \times 10^{-2}}$$

We can perform the calculation as follows:

$$\left(\frac{4.3 \times 5.0 \times 1.29}{2.23}\right) \times \left(\frac{10^{-4} \times 10^{2} \times 10^{-1}}{10^{-2}}\right)$$

Using the rules discussed earlier, it is straightforward to show that

$$\frac{10^{-4} \times 10^{2} \times 10^{-1}}{10^{-2}} = 10^{-1}$$

Let us now make the following approximations, $4.3 \approx 4$, $1.29 \approx 1$ and $2.23 \approx 2$. With these approximations we can write

$$\frac{4.3 \times 5.0 \times 1.29}{2.23} \approx \frac{4 \times 5 \times 1}{2} = 10$$

Thus, an estimate for the calculation is

$$10 \times 10^{-1} = 1$$

This compares with the exact answer,

1.244

With practice an estimate for a complicated calculation can be reached quickly. This is sometimes a useful option to consider in order to ensure a particular calculation will give a reasonable answer without the necessity of working out the exact answer.

Note that similar comments can be made when dealing with scientific notation to be discussed in Section 1.3.

The Worked Example 1.5 shows an important use of standard notation that is not often emphasized, namely the ease with which we can quickly estimate the answer to a problem.

## 1.3   Scientific notation

Scientific notation is closely related to standard notation in that it is based upon powers of 10 to represent numbers. In scientific notation we assign certain values of the exponent a unique name and symbol. Selected prefixes are given in Table 1.4. Note that the exponents are arranged in multiples of 3. It is also a general rule that the positive exponents of 10 are represented by upper-case letters while the negative exponents are represented by lower-case letters. There is one exception to this general rule, namely where we use the symbol 'k' for kilo.

■■■■ **WORKED EXAMPLE 1.6**

Express the numbers given in Worked Example 1.3 in scientific notation.
− Scientific notation suggests we write each of the numbers in the form

$$a \times 10^b$$

where now there is no restriction on $a$ but the exponent is chosen from one of the values given in Table 1.4. We find

(a) $52.9 \, 10^{-12}$ m = 52.9 pm,    (b) $0.299\,792\,458 \, 10^9$ m s$^{-1}$ = 0.299 792 458 Gm s$^{-1}$.

Note that we could also express the result (a) in various ways, for example as 0.052 9 nm − in fact this is how this particular number, the radius of the hydrogen atom, is usually quoted. Similarly, we could express the result (b) as, for example, 299.792 458 Mm s$^{-1}$. It is generally advisable to try to choose the value of the exponent such that the multiplying factor has a value close to unity if this is convenient.

*Table 1.4*

| Multiplying factor | Prefix name | Symbol |
|---|---|---|
| $10^{12}$ | tera | T |
| $10^9$ | giga | G |
| $10^6$ | mega | M |
| $10^3$ | kilo | k |
| $10^{-3}$ | milli | m |
| $10^{-6}$ | micro | μ |
| $10^{-9}$ | nano | n |
| $10^{-12}$ | pico | p |

Often in a problem we will not be given quantities of interest in SI units. Our first step in solving the problem will be to convert all the quantities given to SI units. The advantage of using SI units in an equation is that we know that the answer we obtain will be in its respective SI unit. Worked Example 1.7 shows how we can convert some simple quantities into their respective SI units.

━━━ **WORKED EXAMPLE 1.7**

Convert the following to quantities to their respective SI units:

      (a) a length of 1 mm,    (b) an area of 1 mm².

− We first note that the SI units of length and area are the metre (m) and the square metre (m²), respectively. We find

(a) Making use of Table 1.4 we have 1 mm = $10^{-3}$ m.

(b) To convert mm² to m² we first note that one square millimetre represents the area of a square of side 1 mm so that

    1 mm² = 1 mm × 1 mm

Making use of part (a) we can write

    1 mm² = $10^{-3}$ m × $10^{-3}$ m

Now making use of the rule (1.3) we have

    1 mm² = $10^{-6}$ m²

---

We generally only make use of the SI base units and those derived from the base units. There are only three exceptions to this general use. The first is in the use of the degree Celsius (°C) for the use of temperature as distinct from the strict SI unit of the kelvin (K). In connection with this, we note that a *change* in temperature of one kelvin is equal to a change in temperature of one degree Celsius (°C) and since we are most often interested in temperature changes, we sometimes take the unit of temperature as the degree Celsius. The second special case that we will sometimes encounter in this text is the use of the centimetre (cm) as a unit of measurement. We have

    1 cm = $10^{-2}$ m

The centimetre is not strictly an acceptable SI unit because its relationship with the metre does not make use of one of the selected prefixes given in Table 1.4. The third special case is the use of the degree (°) to measure angles. The strict SI unit of angular measure is the radian. The relationship between the radian and the degree is discussed in the Appendix.

━━━ **Problems**

**1.1**   We are given that the relationship between some quantities $X$ and $Y$ are defined by the equations

(a) $X = mv^2/r$,   (b) $Y = 6v/r$,

where the quantity $m$ has the units of mass, $v$ has the units of speed and $r$ has the units of length. Determine the appropriate SI units of $X$ and $Y$.

*Hint*: Make use of the fact that a number, such as '6' in (b), does not have any units.

**1.2**   Making use of Table 1.3, show that $Q \times V$ has the same units as energy, where $Q$ has the units of electric charge and $V$ has the units of electrical potential.

**1.3**   Convert the following to standard notation:

(a) 593.2, (b) 0.004 287, (c) 215 mA, (d) 85 µA, (e) 5 000 mV, (f) 2 000 µF.

**1.4**   Write the following quantities in scientific notation:

(a) $2\,500\ \Omega$, (b) $7\,500\,000\ \Omega$, (c) 0.000 075 J, (d) 750 000 V, (e) 0.000 000 000 047 F.

**1.5**   Perform the appropriate mathematical operations on the following examples and put in standard and scientific notation:

(a)   $5200 \times 2.6\ 10^{-7}$,   (b)   $6.8 \times 10^4/(9.2 \times 10^3)$,   (c)   $2.5 \times 10^2 + 7.9 \times 10^3$,

(d) $2.5 \times 10^4 - 7.9 \times 10^3$.

**1.6**   A container has a volume of $50\ \text{mm}^3$. Express the volume in terms of the SI unit of volume, namely the cubic metre, i.e. $\text{m}^3$.

**1.7**   Confirm that the following are correct:

(a) $1\ \text{m}^2 = 10^4\ \text{cm}^2$, (b) $1\ \text{m}^3 = 10^6\ \text{cm}^3$.

# CHAPTER TWO

# An introduction to electric charge

Electricity may be described as the phenomenon of electric charge in motion. Any study of electricity, and the closely related phenomenon of magnetism, is necessarily concerned with the study of the behaviour of electric charges. To better appreciate and understand even the simplest of electric circuits it is first necessary to discuss what is meant by 'electric charge'.

In this chapter we introduce electric charge as something which explains the results of some simple experiments. We will be interested in how charges interact with each other, i.e. in the forces that exist between them. One of our main conclusions will be that electric charges interact with each other under the influence of *electric forces*.

Having discussed the effects of electric charge, it is then convenient to discuss the *source* of electric charge. We will find that all of matter is made up of simple components called atoms, which are made up of equal amounts of positive and negative charge.

## 2.1 Electric charge

It is difficult to define exactly what electric charge is. Rather, we try to understand the concept by becoming familiar with the *effects* electric charges produce. Thus, if we rub certain materials with a cloth, it is found experimentally that such materials either attract or repel other similarly rubbed materials. To 'explain' this experimental fact we introduce something called 'electric charge' and attribute certain properties to this quantity.

Let us consider the following experiment: Two strips of polythene are rubbed in turn with a dry cloth and it is arranged that one of the strips of polythene hangs freely from a piece of nylon thread. This ensures that the polythene will be able to respond to any forces exerted upon it. When the second rubbed polythene strip is brought close to the suspended strip there is a force of repulsion, i.e. there is a force between the polythene strips that tries to push them apart; see Figure 2.1.

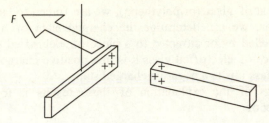

*Figure 2.1*   **Repulsion between two strips of polythene when rubbed with a piece of cloth**

This behaviour is in sharp contrast with the situation before the strips of polythene were rubbed, when no force between the strips is observed. Clearly, something has happened to the polythene. The beginning of the 'explanation' is to argue that the action of rubbing the polythene leaves the polythene *electrified* or in a *charged state*; originally the polythene was in an *uncharged state* (the reasons for the '+' signs in Figure 2.1 will be explained later). The force between the two pieces of polythene is, then, to be 'explained' as some interaction between the charged states of each piece of polythene.

Similarly, if one of the pieces of electrified polythene is replaced by a piece of glass which has been rubbed with a piece of cloth, it is again found that there is a force between the materials that tries to push them apart. Evidently, when glass is rubbed with a piece of cloth it is left in the same type of charged state as polythene when it was rubbed with a piece of cloth.

Finally, let us consider an experiment in which two pieces of plastic are rubbed with a piece of cloth. Again, it is found that the two pieces of plastic try to push each other apart and, hence, it can be argued that the pieces of plastic have been left in a charged state. Now, suppose that we replace one of the pieces of plastic with a piece of electrified polythene. Experimentally, there is a force between the materials – this much is expected – but now the force is such that the materials attract each other, i.e. the force acts between the polythene strip and the plastic and tries to pull them towards each other. Evidently, the plastic is left in a charged state which is somehow different from that of the polythene and the glass.

These, and numerous other experiments, suggest that there are only two types of charged state, namely either the same charged state as that of the electrified polythene or of the electrified plastic. These two charged states may be conveniently labelled 'positive' and 'negative'. These labels really just tell us that the two charged states are different in some manner. We could label these states in any convenient way, for example by 'black' and 'white' or '0' and '1'. When we meet Coulomb's law, the 'positive' and 'negative' labels are found to be the most convenient mathematically.

Purely by convention, the charged state acquired by a piece of glass is called positive. When we rub a piece of polythene, this charged state is indistinguishable from that of the charged state of a piece of electrified glass, so we are forced to label polythene's charge state as positive. Since the charged state of plastic is different from that of glass (or polythene), we are forced to label this charged state as negative. Thus, we can determine the charged state of a body by observing whether it is repelled by or attracted to a piece of electrified glass. If the body is repelled by a piece of electrified glass it is in a positive charged state, while if it is attracted to the glass it is in a negative charged state.

The beginnings of the explanation of these results is to introduce the very important concept of

> electric charge

For the moment 'electric charge' is a different term for 'charged state'. There being two charged states means that we must introduce two types of electric charge, namely positive and negative. A charged or electrified state of a material is explained by the material having acquired, by some means, a quantity of either positive or negative electric charge, and the force between two electrified bodies is interpreted in terms of *electric forces* acting between the charges.

The study of stationary electric charges is referred to as electrostatics. The experiments described above can be 'explained' by assuming that, for some reason, electric charge is such that

> like charges repel; unlike charges attract

The above law is very important and is known as the *First Law of Electrostatics*.

Note that we have not really explained what electric charge is. We introduced it to explain a series of experimental results. We have to do more than just introduce the term 'electric charge', we have to attribute to it certain properties, such as by arguing that like charges repel and unlike charges attract, in order to explain the results of these experiments. At its most fundamental level we cannot answer the question 'What is electric charge?'. This should not worry the reader unduly since we could make similar comments regarding, for example, mass, yet mass is a property that we feel reasonably comfortable with. The reason for this is that we are aware of its consequences through everyday experience. For example, if it requires a large force to make an object move we naturally associate with the object a large mass. Similarly, we can gain some sort of understanding of the concept of electric charge by becoming more familiar with its consequences.

━━━ **WORKED EXAMPLE 2.1**

What is the charged state of a body if it is attracted to a piece of electrified glass?
Since the body is attracted to the electrified glass it must be in a negatively charged state and has, therefore, acquired some amount of negative electric charge.

In the following it will be necessary to consider the amount of electric charge present. We will generally denote the amount of charge a body has by the symbols

> $Q$ or $q$

The symbol $Q$ or $q$ representing the amount of electric charge will have both a magnitude and a sign: the magnitude represents the amount of charge, while the sign (positive or negative) tells us whether the body will be repelled by or attracted to a piece of electrified glass.

The SI unit of electric charge is the coulomb, C. This represents a specific amount of electric charge on a body in the same way that the kilogram represents a specific amount of matter contained in a body.

## 2.2 Coulomb's law and the force between electric charges

We explained the force between two electrified bodies in terms of the force between something called electric charge, arguing that there was a force of repulsion between electric charges of the same polarity (the polarity of the charge means the sign of the charge, i.e. either both positive or both negative) and of attraction between electric charges of the opposite polarity. The law that governs the size of the force between two charges was investigated by the French scientist Coulomb. In a series of careful experiments, Coulomb measured the electric force, $F_E$, acting between two point charges $Q_1$ and $Q_2$ separated by a distance $r$, as shown in Figure 2.2. When Coulomb kept the charges constant, and varied only the charge separation $r$, he found that the magnitude of the force between the charges changed according to an inverse square law. This is written mathematically as

$$F_E \propto \frac{1}{r^2}$$

and is to be read as

'the electric force is proportional to the inverse of the square of the separation'.

Thus, if the force between two charges takes some value, say, $F_E$, when the charge separation is $r$, when the charge separation is doubled, the force acting on the charges decreases by a factor of four (this follows since $1/2^2 = 1/4$). Similarly, if the separation is increased by a factor of three, the force decreases by a factor of nine. The concept of proportionality is discussed in the Appendix.

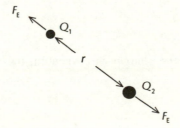

*Figure 2.2*   **Coulomb's law**

Keeping the charge separation constant, and varying the magnitudes of the charges $Q_1$ and $Q_2$, Coulomb found that the force acting on each charge was proportional to the product of the magnitudes of the charges, i.e.

$$F \propto Q_1 \times Q_2$$

so that if, say, $Q_1$ is kept fixed in value, doubling the charge $Q_2$ increases the force acting on both charges by a factor of two. Coulomb concluded that the electric force between two point charges is given as

$$F_E \propto \frac{Q_1 \times Q_2}{r^2}$$

In this form, Coulomb's law can only give the *relative* size of forces. We cannot, yet, determine the *absolute* size of the force. The above *proportionality* is turned into an *equality* by introducing a constant $K$ and writing

$$F_E = K \frac{Q_1 \times Q_2}{r^2} \tag{2.1}$$

where the constant $K$ depends on the medium separating the charges and must be found from experiment. The constant $K$ relates mechanical quantities, i.e. the force $F_E$ and the separation $r$, to electrical quantities, i.e. electrical charge. We shall refer to Equation (2.1) as Coulomb's law.

We now derive appropriate units for the constant $K$. For this we make use of the defining equation (2.1) and note that the units on both sides of this equation must be the same. This means that we must choose the units of $K$ such that when $K$ is multiplied by $C^2$ (the unit of the charge $Q_1$ multiplied by the charge $Q_2$) and divided by $m^2$ (the unit of the separation $r$ multiplied by itself) the result is in newtons (the SI unit of force). Equation (2.1) can be written in terms of units in the following manner:

$$F_E \, [\text{N}] = K \, [?] \, \frac{Q_1 \, [\text{C}] \times Q_2 \, [\text{C}]}{r^2 \, [\text{m}^2]}$$

since the SI units of force and distance are the newton (N) and metre (m), respectively, and we are given that the SI unit of charge is the coulomb (C); see Section 1.1. It follows that a suitable SI unit for $K$ is

$$\frac{\text{N m}^2}{\text{C}^2} \quad \text{or} \quad \text{N m}^2 \text{C}^{-2}$$

If the medium separating the charges is a vacuum, the value of $K$ as determined from experiment is

$$K = 9 \times 10^9 \, \frac{\text{N m}^2}{\text{C}^2}$$

The constant $K$ determines the 'ability' or 'ease' with which electric forces are transmitted between electric charges in free space.

We note that force is an example of a *vector* quantity, i.e. it has both a *magnitude* and a *direction*. The magnitude gives us information about the strength of the force, while the direction tells us the direction in which a body under the influence of the force will move. So far, we have discussed only the magnitude of the electric force; to completely specify the force we must mention the direction in which this force acts. Coulomb showed that the force is directed along the line joining the two charges, as shown in Figure 2.2. Technically, the electric force is an example of a central force, but this is not important for our discussion.

We adopt the convention that if the force, $F_E$, between two point charges is positive then the charges repel each other (i.e. tend to move away from each other), while if the force is negative the charges attract each other (i.e. tend to move towards each other). From this observation we note that if $Q_1$ and $Q_2$ are of the same sign (i.e. $Q_1$ and $Q_2$ are both positive charges or are both negative charges), then the product $Q_1 \times Q_2$ is positive, the force is therefore positive and, hence, the charges repel each other. On the other hand, if $Q_1$ and $Q_2$ are of the opposite sign (i.e. $Q_1$ is positively charged and $Q_2$ is negatively charged or vice versa), the product $Q_1 \times Q_2$ is negative, the force is therefore negative, and the charges attract each other.

Figure 2.3 shows how the force between two charges varies with the separation, $r$, between the charges. The figure illustrates clearly the meaning of the term 'inverse square law'. Note that as we increase the separation the force between the charges gets smaller. In particular, as the separation is doubled the force falls off by a factor of four. Similarly, when the separation is increased by a factor of three the force is only one-ninth of its original value.

For completeness, we point out that in advanced electrostatics the constant $K$ is often written in the form

$$K = \frac{1}{4\pi\varepsilon_0}$$

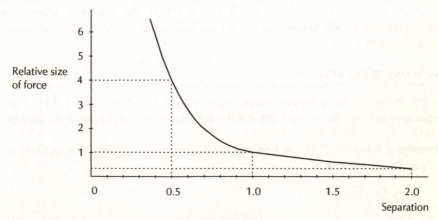

*Figure 2.3* **The meaning of the inverse square law**

where the term $\varepsilon_0$ is referred to as the 'permittivity of free space' and has a value given by

$$\varepsilon_0 = 8.854 \times 10^{-12} \; \frac{C^2}{N\,m^2}$$

Writing $K$ in this form has the advantage that it simplifies other equations that are used more extensively than Coulomb's law in advanced electrostatics.

Worked Examples 2.2–2.4 involve simple manipulation of the Coulomb force law (2.1).

━━━ **WORKED EXAMPLE 2.2**

What is the electric force between two point charges of 3 mC and $-12$ mC placed a distance of 6 mm apart?

— According to Coulomb's law, (2.1), the force is given as

$$F_E = (9 \times 10^9) \times \frac{(3 \times 10^{-3}) \times (-12 \times 10^{-3})}{(6 \times 10^{-3})^2} \; N$$

which comes out to be $-9 \times 10^9$ N. The minus sign indicates that the force between the two charges is attractive, i.e. the charges tend to move towards each other.

Notice that the first step in this calculation was to convert all the numbers into SI units, i.e. to write the charges 3 mC as $3 \times 10^{-3}$ C and $-12$ mC as $-12 \times 10^{-3}$ C, respectively, and the separation 6 mm as $6 \times 10^{-3}$ m. It then follows that the answer, the force between the charges, will be given directly in newtons.

Before progressing further let us consider the size of the force acting between the charges found in Worked Example 2.2. The force between two charges of the order of milli-coulomb separated by a distance of the order of several millimetres turned out to be roughly $10^{10}$ N. To appreciate this figure note that the weight of a medium-sized apple is roughly one newton. It follows that the coulomb represents an enormous amount of electrical charge and we would not expect to deal with charges of such size in practice.

━━━ **WORKED EXAMPLE 2.3**

Given that the electric force between two charges of 4 mC and 8 mC is $5 \times 10^8$ N, determine the separation between the charges. What is the force between the charges if the separation is doubled?

— Rearranging the result (2.1), we find that we can write the separation, $r$, in terms of the force $F_E$ and the charges $Q_1$ and $Q_2$ as

$$r^2 = K \frac{Q_1 \times Q_2}{F_E}$$

from which, upon substituting for the force, the charges and the constant K, we find

$r = 24$ mm

If the separation is doubled, the force decreases by a factor of four. This follows since the force law depends inversely upon the square of the separation of the charges, and is

$$\frac{\text{size of force at } r = 24 \text{ mm}}{4} = \frac{5 \times 10^8 \text{ N}}{4} = 1.25 \times 10^8 \text{ N}$$

In the second part of the example it was straightforward to make use of the fact that the electric force between the charges varies as the inverse of the square of the separation. The alternative is to substitute $r = 2 \times 24$ mm $= 48$ mm into the force law and to determine the force as in Worked Example 2.2.

■■■■ **WORKED EXAMPLE 2.4**

The charge $Q_1 = 2$ mC is separated by a distance $r = 16$ mm from another charge $Q_2$. The electric force between the charges is $4 \times 10^8$ N and acts such that the two charges attract each other. Determine both the magnitude and the sign of the charge $Q_2$.
− If we know the force $F_E$, the charge $Q_1$ and the separation of the charges, $r$, then we can rearrange Equation (1.1) to find that the unknown charge $Q_2$ is given as

$$Q_2 = \frac{1}{K} \frac{F_E \times r^2}{Q_1}$$

From this result, using $F_E = -4 \times 10^8$ N (the minus sign indicates that the force is attractive), we find that $Q_2$ is $-5.7 \times 10^{-3}$ C ($= -5.7$ mC).

## 2.3 Electric charge and matter

It is now convenient to discuss the ultimate source of electric charge, namely matter. We will not delve too deeply into this subject, being content merely to be able to understand the experimental results discussed in Section 2.1. This discussion will also help us to develop an understanding of the conduction process in metals and the difference between a conductor and an insulator.

All matter is composed of atoms. In turn, an atom is composed of three types of particle: the electron, the proton and the neutron. These particles can be distinguished from each other by their mass and their charge, as shown in Table 2.1. We note that the electron and proton are both charged, and that the magnitude of the electron charge is exactly equal to that of the proton but they have opposite signs. Because of our convention of calling electrified glass positive, it turns out that we have to assign the proton charge as positive and the electron charge as negative (this will be explained in Section 2.4). The neutron is an electrically neutral particle and, as such,

*Table 2.1*

| Particle | Mass (kg) | Charge (C) |
|---|---|---|
| electron | $9.11 \times 10^{-31}$ | $-1.60 \times 10^{-19}$ |
| proton | $1.67 \times 10^{-27}$ | $+1.60 \times 10^{-19}$ |
| neutron | $1.67 \times 10^{-27}$ | 0 |

we will not have much need to refer to it, since it plays no role in the electrical concepts to be discussed. Note that the electron is much less massive than either the proton or the neutron.

The simplest atom we can discuss is that of hydrogen (chemical symbol H), which consists of a single proton and an electron. A particularly useful model which will be of enormous help to us is known as the Bohr–Rutherford model of the hydrogen atom, in which the electron moves around the proton in a well defined circular orbit (see Figure 2.4).

Bohr and Rutherford visualized the atom as being constructed like a miniature solar system in which the electron orbits the stationary proton like a planet moving around the Sun. In this model it is the electric force of attraction between the (negatively charged) electron and the (positively charged) proton that keeps the electron bound to the proton. There are several important points to note:

- Since the charges on the electron and proton are exactly equal and opposite, the hydrogen atom is electrically neutral.
- The proton is approximately 2000 times more massive than the electron, therefore most of the mass of the hydrogen atom is concentrated in the proton.
- The proton is sometimes referred to as the *nucleus* of the hydrogen atom and is of size approximately $10^{-15}$ m, and is, therefore, very much smaller than the hydrogen atom, which is of size approximately $10^{-10}$ m. Hence, most of the mass of the hydrogen atom is concentrated in a very small volume, i.e. the nucleus. To try and appreciate these sizes, imagine the atom expanded to the size of the Earth. The proton would then be about 50 m in diameter with the electron orbiting some 6500 km away.

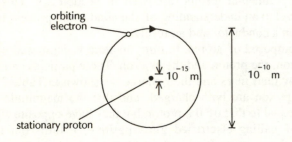

*Figure 2.4*  **The hydrogen atom**

- Unlike a planet revolving around the Sun, which is confined to move in a plane, the electron can be thought of as tracing out a shell. The effect of this is that the hydrogen atom acts like a small sphere.
- The electron can only orbit at certain distances from the proton. This is unlike the case of a planet which can orbit at any distance from the Sun. This introduces the concept of 'allowed shells'.
- Any atom in its normal state is electrically neutral. It therefore follows that the atom has as many orbiting electrons as there are protons in its nucleus. For the case of hydrogen, the nucleus consists of a single proton, so that the hydrogen atom must have one electron orbiting it.
- Since the atom is held together by electric forces, the size of the constant $K$ (or of $\varepsilon_0$) appearing in Coulomb's law (Equation 2.1) determines the size of the atom. For example, making $K$ larger implies that the electric forces holding the atom together are stronger, which in turn implies that the electrons are held more tightly to the nucleus, which implies that the atom would be smaller.

The Worked Example 2.5 makes use of the Coulomb force law to determine the speed with which the electron moves when orbiting the proton.

○ ▬▬▬ **WORKED EXAMPLE 2.5**

Determine the speed with which the electron moves in the hydrogen atom if the radius of the orbit is $r = 5.29 \times 10^{-11}$ m.

− Suppose the electron travels in a circular path of radius $r$ and with speed $v$. If the mass of the electron is $m$ the centripetal force towards the proton is

$$m\,\frac{v^2}{r}$$

This force is provided by the electrical attraction between the electron and the proton, which is

$$K\,\frac{e^2}{r^2}$$

where we have taken into account that the force is attractive and have dropped the minus sign. Equating these two expressions we find that the speed $v$ may be determined from the result

$$v = \sqrt{\frac{Ke^2}{mr}}$$

Substituting the values for $K$, $e$, $m$ and $r$, we find that the speed $v$ comes out as

$$v = 2.2 \times 10^6 \ \text{m s}^{-1}$$

Let us now consider a more complicated atom, copper (chemical symbol Cu), which consists of the following parts:

the nucleus and the atomic electrons.

As in the hydrogen atom, the nucleus is positively charged, contains most of the mass of the atom and is of size approximately $10^{-15}$ m. The copper nucleus consists of 29 protons and 34 neutrons, and therefore has a charge of $29 \times (1.6 \times 10^{-19}$ C$) = 46.4 \times 10^{-19}$ C. This follows since the charge on each proton is $1.6 \times 10^{-19}$ C and the neutron is neutral. The atomic electrons occupy a space of the order $10^{-10}$ m, so that the copper atom is about the same size as the hydrogen atom.

In their usual state atoms are electrically neutral, so that it follows that, since the nucleus of the copper atom consists of 29 protons, then it must also have 29 electrons. This result is emphasized in Table 2.2. Just as for the hydrogen atom, the electrons in the copper atom move around the nucleus only at certain distances, i.e. the electrons are arranged in shells around the nucleus. Each shell can only accommodate a certain number of electrons. Only a maximum of two electrons can occupy the innermost shell (referred to as the K shell), while the next shell (the L shell) is fully occupied when it contains eight electrons. The next shell (the M shell) can contain up to a maximum of 18 electrons, while the outermost shell (the N shell) can contain at most 32 electrons. When the electrons begin to fill up these shells, they do so from the innermost shell outwards. Thus, the innermost shell can only accept a maximum of two electrons, which leaves the remaining 27 electrons to fill the remaining L, M and N shells. Since the L and the M shells can accommodate a total of 26 electrons, the remaining electron must occupy the N shell. How the electrons are distributed between the different shells is referred to as its electronic structure. In the Bohr–Rutherford model the copper atom has the structure depicted in Figure 2.5.

Note that the K shell electrons are the most tightly bound, since they are closest to the nucleus and, hence, feel a larger attractive force. The L shell electrons are not quite as tightly bound as the K shell electrons. This follows because, not only are the L shell electrons further away from the positively charged nucleus, but also the K shell electrons shield the effects of the positively charged nucleus, effectively reducing the amount of positive charge felt by these electrons. Similarly, the M shell electrons are even less tightly bound to the nucleus, and, finally, the N shell electron is the least tightly bound to the nucleus. For copper, it turns out that the outer

*Table 2.2*

|  | Charge (C) | Charge (units of e) |
|---|---|---|
| nucleus | $+46.4 \times 10^{-19}$ | $+29$ |
| orbiting electrons | $-46.4 \times 10^{-19}$ | $-29$ |
| atom as whole | $0$ | $0$ |

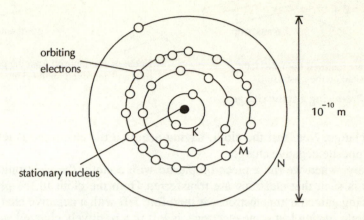

orbiting electrons

stationary nucleus

$10^{-10}$ m

K
L
M
N

*Figure 2.5*   **The copper atom**

electron is only very weakly bound to the atom and only a small amount of energy is required to remove it. The electrical properties of copper are determined by this outer electron, as we shall see in Chapter 3.

## 2.4  Explanation of the results

We are now in a position to explain the experiments discussed at the beginning of this chapter. In normal circumstances, materials are in an uncharged state, being made up of equal amounts of positive and negative electric charge.

An important point to note from our discussion of the atom is that it is much easier to add or remove electrons from an atom than it is to add or remove protons. Indeed, all phenomena of static electricity can be explained in terms of the transfer of electrons only. There are two main reasons for this:

- electrons form the outermost part of the atom, while protons form part of the nucleus which is located at the centre of the atom; and
- electrons are very much less massive than protons and so respond more readily to any forces applied to them than do protons.

It turns out that when polythene is rubbed with a cloth the electronic structure of polythene is such that electrons are readily removed. Choosing the charge of electrified polythene to be positive means we must assign electrons to have negative charge. This follows since the polythene was originally electrically neutral and the effect of removing negative charge will be to leave the polythene positively charged. Since the proton has opposite charge to the electron, we must assign protons to have

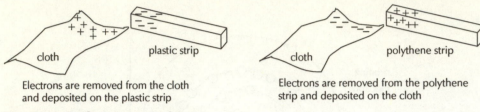

Electrons are removed from the cloth
and deposited on the plastic strip

Electrons are removed from the polythene
strip and deposited on the cloth

*Figure 2.6*   **Producing a static charge**

a positive charge. Note that the cloth, having acquired the electrons, is left with an equal but opposite negative charge.

Conversely, when we rub a piece of plastic with a cloth, the electronic structure of plastics is such that electrons are transferred from the cloth to the plastic. The plastic, having gained extra electrons, is therefore left with a negative charge. In this case, the cloth, having lost some electrons, is left in a positively charged state.

Note that electric charge is conserved overall, i.e. when the polythene loses electrons to the cloth, it is left in a positively charged state, while the cloth is left in a negatively charged state since it has gained electrons. The total amount of electric charge is the same before and after; it has just been distributed differently. Similarly, when plastic is rubbed with a piece of cloth, the negative charge lost by the cloth is gained by the plastic; see Figure 2.6.

## 2.5   Uses of static charges

Static electricity can be a nuisance, but it also has a number of useful applications:

- Spray painting: A high voltage grid is placed in front of the spray gun. The grid acquires a positive charge and the object to be painted a negative charge. As the droplets of paint pass through the grid, electrons are removed from the paint leaving them positively charged. The positively charged droplets are attracted to the negatively charged object. This process helps to prevent the waste of the paint and also produces a uniform finish.
- Dry photocopy machine: The dry photocopy machine uses an aluminium drum coated with selenium. Selenium is a material whose ability to let electric charge move depends upon the light intensity. This ability of selenium to let charge move is called its conductivity. Thus, when selenium is in the presence of light its conductivity is larger than when in darkness. A high-voltage wire located near the drum causes the selenium to acquire a positive charge as it rotates. The drum is in darkness when it is charged. An image of the material to be copied is allowed to fall on the drum. The light portions of the image cause the selenium to have a locally large conductivity compared with parts of the image which are dark. In regions where the conductivity of selenium is large, negative electrons from the aluminium drum can move easily and neutralize the positive charge on the

selenium at that point. Conversely, in dark parts of the image the conductivity of selenium is small, electric charge cannot move so easily and the selenium retains its positive charge. A dark powder that has acquired a negative charge is applied to the drum. The powder is attracted to the positively charged areas on the drum. The powder on neutral areas falls away. A piece of positively charged paper passes under the drum and attracts the powder from the drum. The paper then passes under a heating element, which melts the powder into the paper, creating a permanent copy of the original.

## 2.6 Comparison of electric and gravitational forces

A force we are generally more familiar with is the gravitational force that acts between two masses. Newton's Law of Gravitation states that the (gravitational) force acting between two particles of masses $m_1$ and $m_2$ separated by a distance $r$ is

$$F_G = -G \frac{m_1 \times m_2}{r^2} \tag{2.2}$$

where the force acts along the line joining the particles and the minus sign tells us that the force is attractive. The constant $G$ is termed the 'gravitational constant' and has a measured value of

$$G = 6.672 \times 10^{-11} \frac{\mathrm{N\,m}^2}{\mathrm{kg}^2}$$

By inspection, the above force law is very similar to the Coulomb force law given in Equation (2.1). In particular, both have the same dependence on separation, i.e. both are inverse square law forces. Indeed, if in the above we replace $m_1$ by $Q_1$, $m_2$ by $Q_2$ and $-G$ by $K$ we retrieve Equation (1.1). Just as we think of $m$ as representing the quantity or amount of matter, we can, by analogy, think of $Q$ as measuring the quantity or amount of charge.

Worked Example 2.6 shows that electric forces dominate gravitational forces when discussing the atom.

Given that we have a 'natural measure' which suggests that the gravitational force is much weaker than the electric force, why is it, then, that it is gravity that seems to dominate our lives? The answer to this can be traced to an important difference between matter and charge. This difference is that, whereas electric charge comes in two quantities, i.e. positive and negative, there is only one type of mass (i.e. there is no negative mass). Also, since opposite charges attract each other, the tendency is for charge to form overall charge neutral states. This inhibits the ability to attract other charge. A definite example makes this clearer: Suppose we have two charges, a positive charge $Q_1$ and a negative charge $-Q_2$. Since the charges are of opposite

━━━ **WORKED EXAMPLE 2.6**

Compare the gravitational force and the electric force between an electron and a proton which are separated by a distance *r*.

− If the masses of the electron and proton are $m_e$ and $m_p$, respectively, the gravitational force, $F_G$, between them is given by Newton's Law of Gravitation,

$$F_G = -G\,\frac{m_e \times m_p}{r^2}$$

where the minus sign indicates that the gravitational force is attractive. The electric force, $F_E$, between the particles is given by Equation (2.1), with $Q_1 = -e$ and $Q_2 = +e$

$$F_E = K\,\frac{(-e) \times (e)}{r^2}$$

We can conveniently compare these two forces by considering the ratio $F_G/F_E$. We find

$$\frac{F_G}{F_E} = \frac{-G\,\dfrac{m_e \times m_p}{r^2}}{-K\,\dfrac{e^2}{r^2}} = \frac{G m_e \times m_p}{K e^2}$$

Note that the separation between the electron and the proton, *r*, is dropped in the final result; hence, this ratio is independent of separation. This is because both force laws depend in the same way on the separation. Substituting for the values of the electron and proton masses and the magnitude of the charge on the electron, we find that this ratio is approximately $4 \times 10^{-38}$. This shows that the gravitational force is very much weaker than the electric force and, hence, gravity can be completely ignored when considering atomic physics.

━━━━━━━━━━━━━━━━━━━━━━━━━━━━━━━━━━━━━━━━━━━━━━━━━━

sign according to the First Law of Electrostatics they will attract each other. When the charges coalesce, the body acts as if it has a charge

$$(Q_1 - Q_2)$$

This new total charge is not as effective as $Q_1$ in attracting further negative charge since it is clearly smaller.

In the case of matter, since matter attracts other matter, the tendency is to attract more matter, i.e. the effect of matter attracting more matter is to build up gravitational effects. Thus a body of mass $M_1$ attracts another body of mass $M_2$ with an (attractive) force given by Equation (2.2). When the bodies coalesce, the total mass of the body is

$$(M_1 + M_2)$$

which is clearly larger than the separate individual masses. It follows that this new body will be more effective in attracting further mass.

## 2.7 Summary

Electric charge was introduced to explain the results of experiments involving the rubbing of certain materials with a piece of cloth. It is found, for example, that when two pieces of polythene are rubbed, there is a force acting between the pieces of polythene which tends to push them apart. Numerous experiments suggested the introduction of two types of electric charge, which can be conveniently labelled 'positive' or 'negative', with the properties that like charges repel each other and unlike charges attract each other. These forces are referred to as 'electric forces'. Thus, electric charges move under the influence of electric forces. Normally objects are electrically neutral, and a charge imbalance results when charge is removed or added to the object, when it is described as 'electrified' or 'left in a charged state'.

We then introduced Coulomb's law, which quantifies these experimental observations, and found that the force acting between two point electric charges depends on the product of the charges and inversely on their separation.

The ultimate source of electric charge is matter and a brief introduction to atomic structure was provided.

In the explanation of the results, it was pointed out that the total amount of electric charge is conserved. When the polythene loses an amount of negative charge (leaving it in a positively charged state), the cloth gains this negative charge, and is left with an equal but opposite amount of charge. Similarly, when plastic is rubbed by a piece of cloth, the negative charge lost by the cloth is gained by the plastic – the cloth, having lost an amount of negative charge is left in a positively charged state, while the plastic has gained an equal but opposite amount of negative charge.

## ▬▬ Problems

For the following, take the constant $K$ that appears in Coulomb's law as $K = 9 \times 10^9 \, \mathrm{N\,m^2\,C^{-2}}$ and the magnitude of the electron charge as $1.6 \times 10^{-19} \, \mathrm{C}$.

**2.1** Complete Table P2.1.

*Table P2.1*

| $Q_1$ | $Q_2$ | Sign of force | Repulsive or attractive? | Charge movement? |
|---|---|---|---|---|
| positive | positive | | | |
| positive | negative | | | |
| negative | positive | | | |
| negative | negative | | | |

**2.2** Make use of Coulomb's force law to determine the force between two charges $Q_1$ and $Q_2$ placed a distance $r$ apart if
(a) $Q_1 = +2$ mC, $Q_2 = +1$ mC, $r = 30$ mm, (b) $Q_1 = +5$ mC, $Q_2 = -2$ mC, $r = 60$ mm,

(c) $Q_1 = -8$ mC,    $Q_2 = +1$ µC,    $r = 10$ mm,    (d) $Q_1 = -2$ µC,    $Q_2 = -1$ mC, $r = 80$ mm.

In each case state whether the force between the charges is attractive or repulsive, and whether the charges tend to move towards or away from each other.

**2.3**    Consider two charges $Q_1 = +5$ µC and $Q_2 = +2$ µC. Determine the electric force, $F_E$, on the charges for the range of separations 20 mm $\leqslant r \leqslant$ 100 mm in 10 mm steps. Plot your results on a piece of graph paper and comment on the shape of the resulting curve.

*Hint:* Make use of the fact that if the charges $Q_1$ and $Q_2$ are fixed, then $F_E \propto 1/r^2$.

**2.4**    Determine the force between the charges $Q_1$ and $Q_2$ in Figure P2.1.

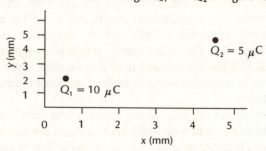

*Figure P2.1*

*Hint:* Determine the distance between $Q_1$ and $Q_2$ and make use of Coulomb's law.

○ **2.5**    Consider the arrangement of three electrical charges, $Q_1$, $Q_2$ and $Q_3$, in Figure P2.2.

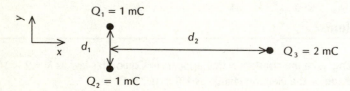

*Figure P2.2*

The distance $d_2$ is kept fixed at 50 mm. Determine the net force acting on the charge $Q_3$ due to the charges $Q_1$ and $Q_2$ if

(a) $d_1 = 50$ mm,       (b) $d_1 = 25$ mm,       (c) $d_1 = 10$ mm,       (d) $d_1 = 1$ mm, (e) $d_1 = 0.1$ mm.

Compare your results with the electric force between two 2 mC charges separated by 50 mm.

*Hint:* Determine the force acting on $Q_3$ due to charge $Q_1$ alone, and due to charge $Q_2$ alone. The total force acting on $Q_3$ is the sum of these two forces, but care has to be taken since they act in different directions. Resolve the force acting the charge $Q_3$ in the $x$ and $y$ directions. The net force acting on $Q_3$ acts along the $x$ direction.

Repeat parts (a), (b), (c) and (d) if $Q_2$ is replaced by a $-1$ mC charge, and comment on your results.

**2.6** (a) A body has charge $+3$ pC. How many electrons have to be added to or removed from the body to attain this charge? (b) Determine the number of protons which have a charge of 1 C.

**2.7** Two particles have electric charges, $Q_1$ and $Q_2$, are arranged so that the second particle may move freely along a vertical rod, as shown in Figure P2.3. If both particles have charge 2 µC and mass 3 mg, determine the height of the second charge above the first. Take the acceleration due to gravity, $g$, as 10.0 m s$^{-2}$.

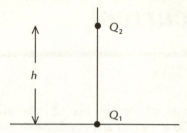

*Figure P2.3*

*Hint:* Determine the forces acting on $Q_2$, namely, the downward force acting on it due to gravity (mg) and the upward force acting on it due to the repulsive electric force (given by Coulomb's law).

o **2.8** Confirm that the gravitational force on a proton on the surface of the Sun (radius $7 \times 10^8$ m and mass $2 \times 10^{30}$ kg) is equal to the electric force between a proton and one microgram of electrons separated by a distance equal to the Sun's radius.

*Hint:* The gravitational force is given by Equation (2.2). To determine the electric force between the proton and the electrons you need to determine the charge possessed by one microgram of electrons. Given the mass of the electron it is straightforward to determine the number of electrons present. Since the charge on each electron is known, it follows that we can determine the required charge.

**2.9** Given the relationship between $K$ and $\varepsilon_0$, confirm the value of $\varepsilon_0$ given in Section 2.2.

# CHAPTER THREE

# *Electric charge in motion: electric current*

We have previously introduced the concept of 'electric charge' to explain our observations concerning the attraction and repulsion of various materials and noted that it came in only two varieties, conveniently labelled 'positive' and 'negative'. We then went on to discuss some basic laws of electrostatics, i.e. laws relating to stationary charges. In particular, we noted a very simple law which applies to electric charge: like charges repel each other, while unlike charges attract each other. This was referred to as the First Law of Electrostatics. After introducing the SI unit of charge, the coulomb, we were able to quantify our observations by making use of Coulomb's law to determine the size of forces acting between two charges.

We now wish to consider moving charges. It will turn out that the simple process of moving electric charge has enormous consequences. In fact, this is what the study of electricity is concerned with – electric charge in motion.

## 3.1 Electric charge in motion

We begin by considering the simple example of many charges all of the same size, say $q$, moving with the same speed and in the same direction past the point P, as indicated in Figure 3.1. Suppose we measure the total amount of charge that passes the point P uniformly in the time $\Delta t$ as $\Delta Q$ (this is simply the number of particles that pass the point P in the time $\Delta t$ multiplied by the charge $q$ of each particle). We then *define* the *electric current* flowing past P as

$$I = \frac{\text{amount of charge passing point P}}{\text{time taken}} = \frac{\Delta Q}{\Delta t} \tag{3.1}$$

Here the symbol $\Delta$ (the Greek symbol *delta*) can be taken to mean 'amount of' or 'change in'. Thus $\Delta t$ is the amount of time that the amount of charge $\Delta Q$ takes in passing the point P.

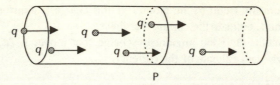

P

*Figure 3.1* **Electricity may be described as the phenomenon of electric charge in motion**

Thus the electric current flowing past some point depends not only on the amount of charge which passes the point, but also how quickly the charge moves past the point. Note that the term 'flow' is really only meaningful when we consider charge to be continuously passing a point, as indicated in Figure 3.1.

### ■ ANALOGY

If we consider fluid flowing past some point, we would naturally define the flow rate of the fluid to be related to the amount of mass $\Delta m$ which passes some point in the time interval $\Delta t$ as the amount of mass that passes the point per unit time, i.e.

$$\frac{\text{amount of fluid passing point P}}{\text{time taken}} = \frac{\Delta m}{\Delta t} \qquad (3.2)$$

We could describe Equation (3.2) as the 'mass current', though a more usual term is the 'rate of fluid flow'. It is a very fruitful analogy to think of electric current flow in terms of fluid flow. If we now replace 'amount of mass passing a point' by 'amount of electric charge passing a point' in the above we retrieve Equation (3.1). Quite naturally we take this 'mass current' to be in the direction in which the fluid moves, and refer to it as the mass or fluid *flow*.

Since the SI units of charge and time are the coulomb (C) and second (s), respectively, this gives the SI unit of current as the coulomb per second (C s$^{-1}$). This is more conveniently referred to as the ampere (A). Thus a current of one ampere is equivalent to a flow rate of electric charge of one coulomb per second:

$$1\,\text{A} \equiv 1\,\text{C s}^{-1}$$

### ■ WORKED EXAMPLE 3.1

The phenomenon of lightning is an example of natural electricity. A charge builds up in clouds that contain a large amount of moisture as they move through the air. The exact details of how this charge builds up are not well understood, but it is generally believed to be associated with charge transfer to rising and falling water crystals in the intense updrafts and downdrafts of the thunderstorm. A typical lightning flash involves the transfer of approximately 25 C of negative charge from the cloud to the ground. If this charge is transferred in a time of the order of a millisecond, estimate the current that flows.

– Using the definition (3.1) the amount of charge that moves is $\Delta Q = 25$ C in the time $\Delta t = 10^{-3}$ s. This gives an estimate of the current as

$$I = \frac{\Delta Q}{\Delta t} = \frac{25 \ C}{10^{-3} \ s} = 25 \ kA$$

We will see later (in particular, see Section 6.5) that when electric charge flows through a material there is an associated heating effect. When lightning occurs the flow of electric current causes a rapid heating of the air, which expands rapidly. The result of this rapid expansion is the thunder we hear which accompanies the lightning.

---

There are two points to be made about this calculation:

■ The transfer involves a flow of *negative* charge. This means that we should use $\Delta Q = -25$ C and the current should be $-25$ kA. The significance of this minus sign will be taken up later.
■ The process of the lightning flash is more complicated than indicated above; the calculation is only meant to estimate the magnitude of the current involved.

■■■ **SPECIAL CASE**

If the current is constant and independent of time, say $I$, then the total charge passing a given point during a time $t$ is given as the product of the current, $I$, and the time that has elapsed, $t$,

$$Q = I \times t \tag{3.3}$$

This result follows simply from the definition (3.1) when the current is constant. We emphasize that this result is only true when the current is constant and does not depend upon time. We will later come across situations when the current depends upon time and we will not be able to make use of the result (3.3).

Conventionally, we use Equation (3.1) to define the coulomb, and argue that the coulomb is that quantity of charge which flows past a point in one second when a

■■■ **WORKED EXAMPLE 3.2**

Determine the amount of charge passing a point in one minute when a constant current of 0.125 A flows.

– Since the current is constant, we can make use of the result (3.3), which gives the amount of charge passing the point as

$$q = 0.125 \ A \times 60 \ s = 7.5 \ C$$

current of 1 ampere flows. We then have the problem of defining the ampere. This will be done when we come to discuss the magnetic effects of an electric current (see Chapter 8).

Thus, when a current of 0.125 A flows, an amount of charge of 7.5 C passes each point in one minute. We will later note that a current of 0.125 A is a quite modest current to flow through an electric circuit. Therefore, although charges of size coulomb are very large in electrostatics, in electric circuits this seems to be a fairly modest amount of electric charge.

### Conventional current flow

As noted previously, there are two types of charge: 'positive' and 'negative'. This leads to some slight, but interesting, complications when we come to use Equation (3.1). We first consider the case when the charge passing some point P is positive, say $\Delta Q > 0$. Since the time this charge takes to pass, $\Delta t$, is always positive, according to our definition (3.1), the current $I$ is found by dividing one positive number, $\Delta Q$, by another, $\Delta t$, and so must be a positive number. We interpret this by saying that the current $I$ flows in the same direction as the charges move see Figure 3.2. This is very similar to the case of fluid flow, and the fact that the (electric) current flow is in the same direction as the (electric) charge movement obviously causes no problems.

Suppose, however, that the charge moving past P is negative, i.e. $\Delta Q < 0$. The time the charge takes to move past the point, $\Delta t$, is still positive, so that using (3.1) to find the current gives us a negative result, since we are dividing a negative quantity by a positive quantity; see Figure 3.3.

How are we to interpret this result? What does a 'negative current' mean? We will interpret a negative current to mean a current flowing in the opposite direction to which it is given. This is similar to interpreting the negative speed of an object as

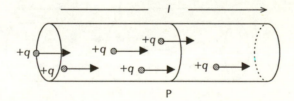

*Figure 3.2* **Conventional current flow and positive charge movement**

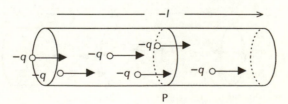

*Figure 3.3* **The electric current and negative charge movement**

taking the object to be moving in the opposite direction assumed. This means that we must take the current associated with the movement of negative charge to flow in the direction opposite to the charge movement; see Figure 3.4.

In the case of the movement of negative charge, this is where the analogy between the movement of electric charge and the movement of a fluid breaks down. The 'mass current' or the rate of fluid flow is always in the same direction that the fluid moves and is easy to visualize. This can be traced to the absence of negative mass, since for a fluid both $\Delta m$ and $\Delta t$ are always positive and we never have to consider what a negative rate of fluid flow means. The meaning of a negative (electric) current flow is more difficult to visualize, since we have charged particles which are moving in one direction, yet we attribute the current to be flowing in the opposite direction.

These results lead us to consider the term 'conventional electric current flow', which may be stated in the following manner:

> the conventional electric current flows in the same direction as the movement of positive charge.

### Some special cases

There are three special cases we can usefully consider:

*Case 1*    Let us consider the case when equal but opposite charges move with the same speed but in opposite directions; see Figure 3.5. From previous arguments, the current associated with the movement of the negative charges flows in the same direction as the current associated with the movement of the positive charges. It follows that these currents add.

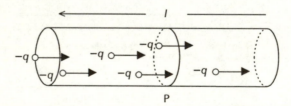

*Figure 3.4*   **Conventional current flow and negative charge movement**

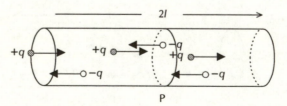

*Figure 3.5*   **Charge movement and current flow: Case 1**

*Case 2*   The result given in case 1 is to be contrasted with that shown in Figure 3.6, which depicts two equal amounts of positive charge moving with the same speed but in opposite directions. In this case there is no net current, since the currents associated with each charged particle cancel each other (a similar conclusion will result if we consider two equal negative charges moving in opposite directions). Admittedly, this is difficult to understand, since if we were to measure the amount of charge that passes the point P we would get a non-zero result. We have to be a bit more careful; rather than just measuring the charge that passes the point P, we also have to take into account the directions in which the different charges are moving.

*Case 3*   Finally, Figure 3.7 shows the case of equal but opposite charges both travelling with the same speed and in the same direction. That no overall current flows is perhaps reasonable in this case, since the total charge passing the point P is zero, so from the definition (3.1) the current should come out as zero. This final result explains why, for example, when we move a piece of, say, metal, there is no flow of electric current – the piece of metal consists of equal amounts of positive and negative charge which are moving in the same direction.

Worked Example 3.3 investigates one consequence of the definition (3.1).

━━━━ **WORKED EXAMPLE 3.3**

Determine the current flowing in a material for which there are $n$ particles per unit volume, when each particle has charge $q$ and is moving with speed $v$, see Figure 3.8.

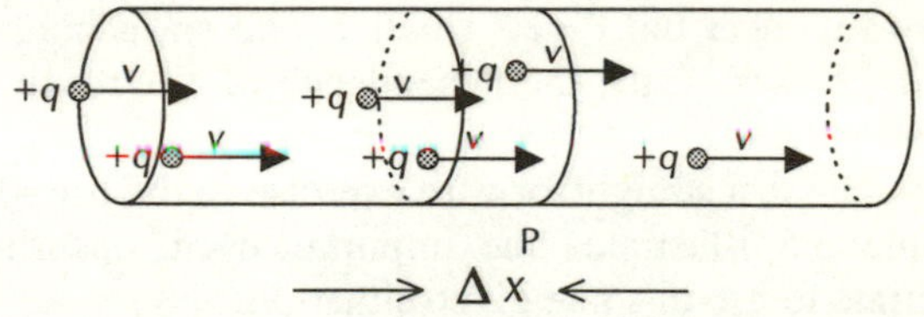

**Figure 3.8   The current flow for a set of charges all moving in the same direction and with the same speed**

– To determine the current flow we merely have to count the charges that move past the point P in a time $\Delta t$. The charge that passes this point is simply given as the product of the number of particles and the charge on each particle, $q$. From Figure 3.8 we note that if the speed of all the particles is $v$, all those particles that are within the distance $\Delta x = v\,\Delta t$ pass through the plane. This number, $N$ say, is equal to the density of the particles, i.e. the number of particles per unit volume, $n$, multiplied by the volume of the element of width $\Delta x$, which is just $A\,\Delta x$, giving

$$N = nA\,\Delta x = nAv\Delta t$$

where $A$ is the cross-sectional area of the material. This gives for the charge passing through the plane during the time interval *t* as

$$\Delta Q = Nq = nvA\,\Delta tq$$

This is the amount of charge which passes through the plane during the time $\Delta t$. It thus follows that the rate of flow of charge across the plane, which is another term for the current, is given as

$$I = \frac{\Delta Q}{\Delta t} = nqvA \tag{3.4}$$

Let us discuss the result (3.4) in a little more detail and consider whether it is reasonable. We note that it is made up of the four factors: *n* (the density of particles), *q* (the charge on each particle), *v* (the speed of each particle) and *A* (the cross-sectional area of the conductor). That the current depends upon the particle density, *n*, is straightforward – this merely tells us that the larger the number of particles that are moving, the greater will be the amount of charge, and the larger will be the current flowing; similarly, the smaller the value of *n*, the smaller the current that will flow when the other factors are kept constant. The dependence upon the size of charge *q* of each particle is also easy to understand; it tells us that if the charge on each particle is larger, then more charge will pass P in the time $\Delta t$. As the cross-sectional area *A* increases so the number of particles that take part in the conduction increases, and hence the amount of charge that passes increases and, therefore, the current increases. Finally, the result (3.4) depends upon the speed of all the particles. We note that as the speed of the charges increases, the time a given charged particle spends near the P gets smaller, and so, according to the definition (3.1), the current increases. Thus, the dependence of the result (3.4) on the factors *n, q, v* and *A* is reasonable.

Worked Example 3.4 is a straightforward exercise in the use of (3.4).

Worked Example 3.5 illustrates the importance of ensuring that a particular formula is appropriate to use in some given situation.

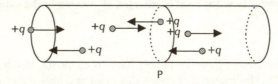

P

*Figure 3.6*   **Charge movement and current flow: Case 2. The net current is zero**

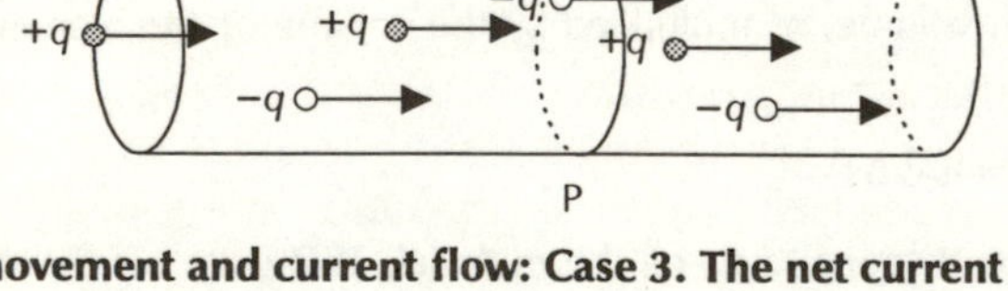

P

*Figure 3.7*   **Charge movement and current flow: Case 3. The net current is zero**

■■■ **WORKED EXAMPLE 3.4**

Given that there are $n = 10^{18}$ particles per cubic metre, each with charge $q = -1.6 \times 10^{-19}$ C, determine the current if all the particles are moving past a plane of cross-sectional area 20 mm² in the same direction with speed $v = 100$ mm s$^{-1}$.

—Making use of the result (3.4), converting all quantities to SI units gives the current as

$$I = (10^{18} \text{ m}^{-3}) \times (-1.6 \times 10^{-19} \text{ C}) \times (20 \times 10^{-6} \text{ m}^2) \times (0.1 \text{ m s}^{-1})$$

giving $I = -0.32 \times 10^{-6}$ A $= -0.32\mu$A. Note that we were able to quote the answer in amperes directly since we converted all the quantities to their appropriate SI units, and the SI unit of current is the ampere. The minus sign indicates that the current is flowing in the direction opposite to the movement of the (negatively charged) particles.

■■■ **WORKED EXAMPLE 3.5**

Consider Worked Example 3.4. Take the charge carriers to be free electrons in a metal which are moving randomly throughout the metal with a mean speed of $10^5$ m s$^{-1}$. Determine the current flowing.

—If we use the result (3.4) it is straightforward to show that the current flowing is given as

$$I = -0.32 \text{ A}$$

However, we should always be careful when we apply a particular result to a particular problem. We should really ensure that the result is appropriate to use in the problem. Is the result (3.4) appropriate to make use of in this problem? The answer is 'No', since an important assumption in the derivation of the result was that the charge carriers are all moving in the same direction. In this example the charge carriers are stated as moving *randomly* with the *mean* speed $10^5$ m s$^{-1}$, i.e. the charges are moving in *different directions* with *different* speeds throughout the material, and the mean speed quoted is some average speed of all the charges. Careful consideration of the derivation of the result (3.4) shows that an assumption was made that the charges were all moving in the same direction. Thus, the result (3.4) does not apply in this problem and we cannot make use of it to determine the current. It is straightforward to show that the current flowing in this situation is actually zero. This follows since the charge carriers are moving randomly, so that at any instant the same number of charges travelling in one direction are also travelling in the opposite direction. Hence, there is no net current flow.

## 3.2 The conduction process

Having defined what we mean by the term 'electric current', it is now appropriate to consider the process of electric conduction in a series of materials. We note that materials are able to conduct an electric current to various degrees and we are able to broadly define two types of material, as shown in Table 3.1.

*Table 3.1*

| Conductors | Insulators |
| --- | --- |
| allows currents to flow easily | inhibit or resist the flow of current |
| most metals … | plastics, rubber … |

We first discuss conductors, the most important of which are metals. This will then enable us to point out the difference between metals and insulators. A class of materials not discussed here is semiconductors. As their name suggests, these materials have electrical properties which lie between those of conductors and insulators. Semiconductors are enormously important in everyday life and will be dealt with in Chapter 14.

## Conductors

Recalling our discussion of the copper atom in Section 2.3, we now introduce some useful notation. When the outermost electron is removed from the copper atom, we are left with a positively charged 'atomic core' or 'ion'. This atomic core consists of the positively charged copper nucleus (with total charge $+29e$, where $e$ is the magnitude of the charge on the electron) surrounded by 28 electrons (with total charge $-28e$). Thus, overall, the charge on the atomic core is $+e$ (i.e. $29e - 28e$); see Figure 3.9. One way of showing that the copper atom has lost an electron is to attach a '$+$' sign to the chemical symbol of copper, i.e. $Cu^+$.

We are now able to present a simple model of the most important class of conductors, namely that of a metal. A metal consists of a regular lattice of these atomic cores, i.e. atoms which have lost one or more of their outer electrons, surrounded by 'free' electrons (Figure 3.10). These electrons are termed 'free' or 'mobile' because they do not belong to any particular atom and are able to move

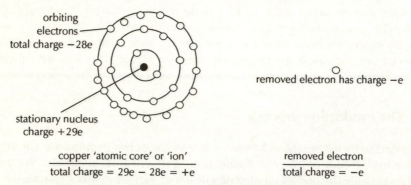

orbiting electrons — total charge $-28e$

removed electron has charge $-e$

stationary nucleus charge $+29e$

| copper 'atomic core' or 'ion' | removed electron |
| --- | --- |
| total charge $= 29e - 28e = +e$ | total charge $= -e$ |

*Figure 3.9*  **Removing an electron from the copper atom leaves it electrically charged**

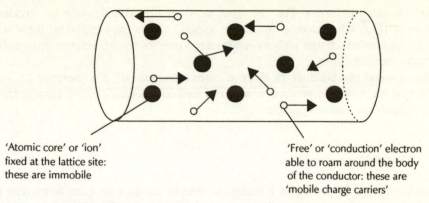

'Atomic core' or 'ion'
fixed at the lattice site:
these are immobile

'Free' or 'conduction' electron
able to roam around the body
of the conductor: these are
'mobile charge carriers'

*Figure 3.10*    **A simple model of a metal**

more or less freely throughout the lattice, although energy has to be supplied for them to leave the surface of the metal. Also, the free electrons are uniformly distributed throughout the metal. The significance of the term 'uniformly distributed' is that if we place a small cube inside the metal, the net charge will, on average, be zero; the amount of positive charge is cancelled by the equal but opposite amount of negative charge.

Let us consider the charge movement in the metal and ask whether any current is flowing. The atomic cores sit in more-or-less fixed positions in the lattice, which are referred to as the lattice sites. At non-zero temperatures, however, they do vibrate about the lattice site. Similarly, we have already mentioned that the (free) electrons are moving randomly throughout the body of the metal. Since we have moving charges, do we have an electric current flowing? The answer is 'No', because at any instant there is no net charge movement past any point in the metal and, therefore, no electric current is flowing. For instance, even though the electrons are moving throughout the lattice, at any particular instant as many electrons are moving to the right, say, as are moving to the left – the electrons are moving randomly through the conductor and there is no net movement of charge. We have defined the electric current in terms of the net amount of charge passing some point in some time. If, on average, no net charge passes the point, then the current flowing is zero (see Worked Example 3.5).

Why do we class a metal as a conductor? Let us consider somehow applying appropriate electric forces to a metal, for example, by bringing an electric charge $Q$ nearby. Both the atomic cores and the free electrons, being charged, feel the electric forces exerted by the charge $Q$ (recall our discussion of the First Law of Electrostatics in Section 2.1 and Coulomb's law in Section 2.2). The atomic cores are very massive and are difficult to move under moderate electric forces (they are sometimes referred to as 'immobile charge carriers'), and hence do not contribute to any flow of electric current. The free electrons, however, are relatively light and since they are able to move freely throughout the lattice, they are able to respond readily to electric forces. For this reason, they are also known as 'conduction electrons'. Thus, when electric forces are applied across the conductor, there will be a net movement of charge and, as emphasized previously, if electric charge moves an electric current flows.

Thus a metal can conduct an electric current because of the presence of free or conduction electrons. In general, a material can conduct an electric current if there are mobile charge carriers present.

## Insulators

A conductor, and in particular a metal, is able to conduct an electric current when electric forces are applied to it because of the presence of mobile charge carriers, i.e. conduction electrons. All we are saying is that if we apply electric forces, then there is electric charge available which is able to move under the influence of such forces. If electric charge moves then, as we have argued, this is another expression for the term 'electric current'.

If we recall Worked Example 3.3, we found the current to depend on the following four factors: $n$ (particle density), $q$ (charge on each particle), $v$ (speed of each particle) and $A$ (cross-sectional area of the material). We can interpret $nq$ as the charge density of the material, call it $\rho$ (the Greek symbol *rho*). A better term for it is the 'mobile charge density', since we have seen that a metal, for example, consists of electric charge which is difficult to move when we apply electric forces (namely the atomic cores), and that it is the amount of charge that is able to move which is important. Thus, Equation (3.4) shows that the current depends upon the amount of mobile charge density available. The basic difference between a conductor and an insulator is really one of only degree. We classify a good conductor as having a large mobile charge density – i.e. a conductor has a large number of charged particles which can move under the influence of electric forces. Conversely, an insulator has only a relatively small mobile charge density – i.e. has only few charged particles that move under the influence of electric forces.

### ▬▬▬ WORKED EXAMPLE 3.6

If, in Worked Example 3.4, there were only $n = 10^{12}$ particles per cubic metre taking part in the charge flow, determine the electric current if all the other factors remain unchanged.
—Making use of the result (3.4), converting all quantities to SI units gives the current as

$$I = (10^{12} \text{ m}^{-3}) \times (-1.6 \times 10^{-19} \text{ C}) \times (20 \times 10^{-6} \text{ m}^2) \times (0.1 \text{ m s}^{-1})$$

from which the current $I$ is given as giving $-0.32 \times 10^{-12}$ A $= -0.32$ pA. The minus sign again indicates that the current is flowing in the opposite direction to the movement of the (negatively charged) particles. But now the current is of the order of pico-amperes and would require sensitive instruments to be measured.

━━━━ **WORKED EXAMPLE 3.7**

A rod of copper $1$ cm$^2$ in cross section and 10 cm long weighs about 90 g. It contains approximately $8.5 \times 10^{23}$ atoms. Suppose that somehow an electron excess of just one part in a million was caused in one half of the rod, with a corresponding deficit in the other. Estimate the force on the charge distribution.

− We begin by determining the total number of electrons and protons in the rod. Since each atom of copper consists of 29 atomic electrons, this means that there are approximately $2.5 \times 10^{25}$ electrons and, therefore, an equal number of protons in the rod. We are interested in the order of the magnitude of the answer, so we shall suppose that the charge distributions can be treated as if they are concentrated at the mid-points of each half of the bar. The problem becomes equivalent to finding the force between a group of $(10^{-6}) \times (2.5 \times 10^{25})$ electrons 5 cm distant from a similar group of protons. This force can be determined from Coulomb's law and is given as

$$F_E = 9 \times 10^9 \times \frac{(-2.5 \times 10^{19} \times 1.6 \times 10^{-19}) \times (2.5 \times 10^{19} \times 1.6 \times 10^{-19})}{(5 \times 10^{-2})^2} \text{ N}$$

which works out to be $-6 \times 10^{13}$ N, the minus sign indicating that the charge distributions attract each other. As it requires only about $10^7$ N to produce severe deformations in the copper rod we can be sure that the electrons will move so as to establish a more uniform distribution. If the electrons were not free to move, the rod would simply collapse.

━━━━━━━━━━━━━━━━━━━━━━━━━━━━━━━━━━━━

Having dealt with electric forces and discussed a simple model of a metal, we backtrack somewhat and provide an interesting example which reflects the strength of electric forces and indicate why it is difficult to achieve charge separation: as soon as we move positive and negative charges apart there are strong restoring forces which try to bring the charges together again. The example is quite detailed, though with some effort, the reader should be able to follow the arguments. It is more important, however, that the sizes of the forces involved be appreciated when the charge equilibrium of a body is disturbed.

This example is also of interest in that it shows the importance of simplifying an otherwise intractable problem in order to gain insight into the order of magnitude of the answer involved. In principle, to solve the problem exactly we would have to calculate all the forces between all of the electrons and between all of the protons. Since there are some $2.5 \times 10^{25}$ electrons and the same number of protons it would clearly be a very difficult problem to determine the forces between all the particles.

━━━━ **COMMENT**

■ Let us try to appreciate what the term 'one part in a million' means. We begin by estimating the size of a football pitch to be 20 m by 70 m, so that the area of the pitch is 140 m$^2$. Now let us imagine covering the pitch completely with red snooker balls. If we take the diameter of a snooker ball as 5 cm this means we would require roughly one half million snooker balls to cover such a pitch. Therefore, we would require roughly one million balls to fill two such pitches side by side. The meaning of the term 'one part in a million' can be fully appreciated if we are asked to imagine replacing just one of these red balls by a ball of a different colour.

o        The electrochemical effect and $I = \Delta Q / \Delta t$

We now discuss the electrochemical effect, i.e. the effect which occurs when an electric current passes through certain fluids. This discussion is merely to emphasize that in the definition (3.1) for the electric current, $I$, the type of charge moving is not important, and the definition applies to *any* charge which is in motion.

Consider Figure 3.11, which shows two plates marked '+' and '−' placed in a copper sulphate solution. For the moment, we imagine the '+' and '−' imply that some positive and negative charge is placed at these positions, respectively. These two plates through which the current enters and leaves the copper sulphate solution are called the electrodes. In this context the copper sulphate solution is referred to as the electrolyte. The electrode at which the current enters the electrolyte is called the anode, while that by which it leaves is called the 'cathode'.

We do not wish to go into all technicalities of the process, but note that copper sulphate is a substance consisting of equal numbers of copper ions, $Cu^{2+}$ (i.e. a copper atom which has lost two electrons), and sulphate ions, $SO_4^{2-}$ (i.e. an $SO_4$ group which has gained two electrons), arranged in a particular manner such that the whole is electrically neutral. When copper sulphate is placed in water, it dissociates into equal numbers of copper and sulphate ions which are able to move freely in the electrolytic solution, i.e. are able to move under the influence of suitable electric forces. In the electrolyte the current is therefore conducted by the movement of both positive and negative charges.

In the following we present a simplified discussion of what occurs in the electrolyte. The $Cu^{2+}$ ions are attracted towards the cathode (the negatively charged plate), while the $SO_4^{2-}$ ions are attracted towards the anode (the positively charged plates). At the cathode, two electrons are given up and we have

$$Cu^{2+} + 2e^- \rightarrow Cu \ (metal)$$

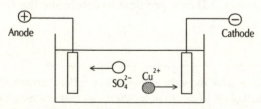

*Figure 3.11*   **Charge movement in an electrolyte**

so that copper metal is deposited there. The reaction at the anode is somewhat more involved; we shall not discuss it except to say that the negative ions give up their electrons there. What is important is the charge movement and the current flow: in the electrolyte the current flow is due to ions, while in the connecting wire it is due to electrons. Notice that in the figure the positive ions travel towards the cathode while the negative ions travel towards the anode. In both cases the current flow is from anode to cathode.

## 3.3  Summary

We introduced the concept of electric charge in motion and referred to this phenomenon as 'electric current', and then proceeded to put this concept on a more formal level, defining the current as the rate of flow of charge, $\Delta Q/\Delta t$. In particular, we noted the problems this definition poses because charge comes in two types: positive and negative. We had to introduce 'conventional current flow' as the direction in which the current flows. The conventional current was taken to flow in the same direction that positive charge moves. The current flow associated with a negative charge is taken in the opposite direction to which the negative charge moves.

We then derived the current flowing for the case of a set of charges all moving with the same speed and in the same direction. Following this we discussed the conduction process in a metal as being due to the presence of mobile charge carriers, i.e. conduction electrons. An insulator was described as an extreme case of a conductor with very few conduction electrons.

Finally, we briefly discussed the charge movement which occurs in an electrolyte. Only a brief description was attempted, the emphasis being that the definition (3.1) applies to *any* charge which is in motion.

## ▬▬ Problems

**3.1**  A piece of material is of cross-sectional area $10^{-2}\,m^2$ and has $10^{18}$ particles per unit volume. The charge on each particle is $-1.6 \times 10^{-19}$ C.
(i) If all the charges move with a speed of 20 mm s$^{-1}$ determine the current *I* flowing.
(ii) If a current of 2 A flows through the material determine the speed *v*.
In both cases comment on the relative signs of *v* and *I*. Comment on the relative signs of *v* and *I* if $q = +1.6 \times 10^{-19}$ C.

**3.2**  In Worked Example 3.1 it was stated that a typical lightning flash involves the (negative) charge transfer of about 25 C. How many electrons are transferred in the flash?

○ **3.3**  Show explicitly that in the formula (3.4)

$I = nqvA$

the right-hand side has the SI unit of electric current.

**3.4**     The copper atom consists of a positively charged nucleus of charge $46.4 \times 10^{-19}$ C. It is surrounded by 29 electrons, each of charge $-1.6 \times 10^{-19}$ C. Explain why there is no electric current when the copper atom moves. Does an electric current flow if a copper *ion* moves?

**3.5**     A constant current of 5 mA flows in an electric circuit. How much charge flows past any point in the circuit in 10 seconds?

**3.6**     The amount of charge passing a point in 5 ms is 250 μC. Determine the current passing the point.

**3.7**     A current of 0.2 A flows through an electric circuit for 10 s. What is the total charge which has passed through the circuit? How many electrons have moved past a given point in the circuit? Take the charge on the electron to be $-1.6 \times 10^{-19}$ C. Discuss the relative directions of the electron movement and the current flow.

# CHAPTER FOUR

# *An introduction to electric circuits*

In Chapter 2 we introduced the concept of electric charge and noted that to explain experiments we had to introduce two types of charge: 'positive' and 'negative'. Coulomb's law was then discussed and we suggested that the experiments could be interpreted in terms of *electric* charges being influenced by *electric* forces. It is important to note that electric charges move under the influence of electric forces.

In Chapter 3 we suggested that electricity could be described as the phenomenon of electric charge in motion. We then discussed the term electric current and defined it formally as the rate of flow of electric charge. Another important point brought out in the discussion was the introduction of conventional current flow, which was taken to flow in the direction that positive charge moves.

We are now in a position to consider a simple electric circuit. In the following we first introduce some terminology which will be important for the rest of the book. After this we discuss an important analogy which will help us understand the roles of the different components of an electric circuit. We then develop our understanding of the role of the battery by noting that ultimately it is the application of *electric* forces which cause *electric* charges to move. Finally, we introduce a link between electric forces doing work on electric charges, forcing charges to move round a circuit and the work done on them. This will enable us to introduce a suitable unit for electromotive force and potential difference. We then comment on the similarities and the differences between the electromotive force and potential difference.

## 4.1 A simple electric circuit – introducing terminology

In this section we introduce some suggestive terminology which will be important for later work. We do this by considering the electric circuit shown in Figure 4.1. Figure 4.1(a) shows a photograph of a real circuit, which consists of a battery, labelled by the symbol $E$, connected by wires to three resistors. Also shown in the figure is a voltmeter, which is used to measure the potential difference *across* one

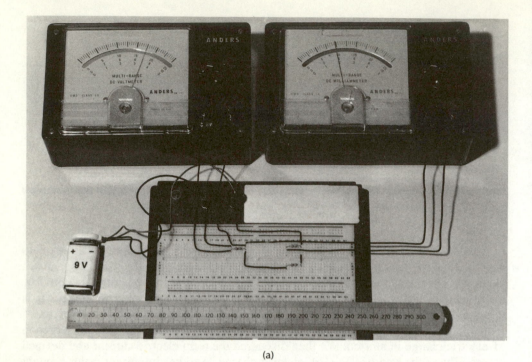

(a)

(b)

*Figure 4.1*   **A simple electric circuit. (a) The real circuit; (b) the circuit diagram**

of the resistors, and an ammeter, which measures the current flowing *through* another of the resistors. Figure 4.1(b) is known as the circuit diagram, which provides a convenient description or model of the real circuit, where generally accepted symbols have been used to represent the different components. The circuit diagram can be thought of as a 'map' of the circuit.

In the circuit diagram, the battery is labelled $E$ and the three resistors are labelled by the general symbol $R$ with subscripts 1, 2 and 3 to distinguish them. The voltmeter is labelled by the symbol V and measures the potential difference across the resistor $R_1$, while the ammeter is labelled by the symbol A and measures the current flowing through $R_2$. The points $J_1$ and $J_2$ are referred to as *junctions*. At the junction $J_1$ the current has a choice of flowing along two different paths or branches, either through the resistor $R_2$ or the resistor $R_3$. The current divides at this junction, the amount of current flowing through each branch depends upon the relative amounts of opposition that $R_2$ and $R_3$ present to the flow of current. At the junction $J_2$ the current from each branch recombines to form the total current that entered the junction $J_1$. The current $I$ that leaves the battery at the positive terminal returns to the battery at the negative terminal.

All we need to note for the moment is that both the voltmeter and the ammeter work by making use of some effect of the electric current, so that as the current increases, for example, some dial will increase proportionally. We note for the present discussion that perfect voltmeters and ammeters do not influence the circuit they are making measurements on.

The battery is said to provide an 'electromotive force' (e.m.f.) of 'amount' or 'size' $E$, which sets up a 'potential difference' (p.d.) across the components in the circuit which drives an 'electric current' through them. Since we have argued that an electric current is merely the movement of electric charge, it follows that the p.d. *across* a component causes electric charge to move *through* the component. These components offer varying degrees of resistance to the flow of current.

## 4.2 A useful analogy

A useful analogy often made with simple electrical circuits is to think in terms of a pipe with some fluid flowing through it due to the action of a pump. We could refer to such a system as a 'fluid circuit'. Figure 4.2 shows a simple fluid circuit consisting of a pump pushing fluid through a simple circuit consisting of two constrictions labelled $C_1$ and $C_2$. Also shown is the analogous electric circuit we wish to understand.

In this system, the pump causes fluid to flow around the pipes by creating a pressure difference between the ends of the pipes. The same amount of fluid that

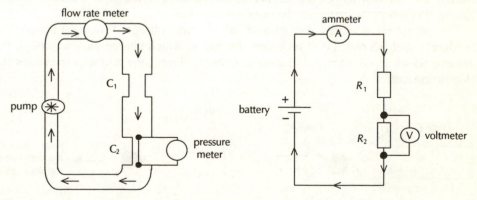

*Figure 4.2* **The analogy of a fluid flowing through a pipe and an electric current**

leaves the pump each second at one end, re-enters the pump at the other end, i.e. the rate at which fluid leaves the pump is equal to the rate at which the fluid re-enters the pump. The size of the pump, or the ability of the pump to move fluid around the circuit, is conveniently measured by the pressure difference the pump is able to apply to the ends of the pipes.

The flow rate meter measures the amount of fluid flowing through a pipe each second, i.e. it measures the rate at which fluid passes through the pipe, say in kilograms of fluid per second or in cubic metres of fluid per second.

The amount of fluid that passes through a given section of pipe depends on the difference in pressure between the ends of that section. Since we are dealing with fluids, this pressure is referred to as 'hydrostatic pressure'. If there is a difference in the (hydrostatic) pressure between the two ends of the section then there will be a net force (which is just pressure times area) acting on the fluid, which will cause it to move in a particular direction. That direction points from, of course, regions of high pressure to regions of lower pressure (see Figure 4.3). The greater this difference, the larger will be the amount of fluid passing through the pipe in each second. This is measured by the pressure meter. Note that the pressure meter does not measure the pressure at a particular point; rather, it measures the *difference* in the pressure between *two* points. The rate at which fluid passes completely around the 'circuit' depends not only on the pressure delivered by the pump, but also on the 'resistance' offered by the pipes to the flow of the fluid. In the fluid circuit, resistance to the flow of fluid is represented by the constrictions in the pipe.

Let us compare this description with that for the electrical circuit. We will find that the descriptions in both cases are very similar.

The battery (in general, the source of e.m.f.) causes mobile charge carriers in the conductor to move around the circuit by creating a difference in 'electrical pressure' between the positive terminal and the negative terminal. The same amount of electrical charge that leaves the battery each second at the positive terminal, re-enters the battery at the negative terminal. But the 'amount of electric charge that passes a point per second' is just the electric current, so the statement is the same as 'the electric current that leaves the battery at the positive terminal is the same as the current that enters the battery at the negative terminal'.

The ammeter measures the amount of electric charge flowing through the conductor each second, i.e. it measures the rate at which charge passes through the system, which is, of course, the electric current. Hence, the ammeter measures the electric current.

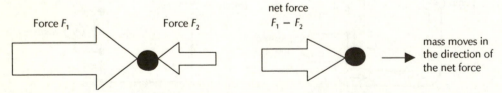

*Figure 4.3*  **An object moves in the direction of the net force acting on it**

The voltmeter measures the difference in the 'electrical pressure' between two ends of a component – the greater the difference in the 'electrical pressure' between the ends of the component, the larger will be the electric force acting upon the electric charges and, therefore, the larger will be the electric current that flows through it.

The rate at which charge passes (or current) completely around the circuit depends not only on the 'electrical pressure' provided by the battery, but also on the resistance offered by the components to the flow of the electric current. Components which offer resistance to the flow of the electric current are referred to as resistors. Resistance will be treated in detail in Chapter 5.

Note that, unless indicated otherwise, the wires connecting together the circuit elements are taken to have zero resistance. Thus, apart from connecting together the various elements which make up the circuit, the wires do not influence the amount of current flowing through the circuit.

Rather than use the term 'difference in electric pressure' we make three steps to arrive at the more common term 'potential difference':

- First we prefer the term 'difference in electric potential' – this makes the connection with electricity.
- This expression is a little unwieldy, so we shorten it to 'electric potential difference'.
- Finally, since we are dealing wholly with electrical circuits the 'electric' part of the term is understood and we shorten the term even further by referring to it as 'potential difference' or 'p.d.'.

That we need a p.d. between two points for charge to move between them (i.e. a current flows) is emphasized by the analogy in Figure 4.4. The conventional current flows from regions of high electric potential to regions of low electric potential, i.e. from the positive terminal of a battery to the negative terminal. Alternatively, the current is said to flow down the electric potential gradient (in the same manner that a particle in a gravitational field can be said to move down the gravitational potential gradient).

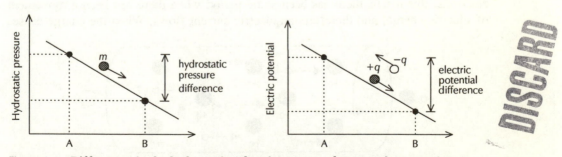

*Figure 4.4*  **Differences in the hydrostatic (electric) pressure between the two points A and B causes fluid (charge) to move between the points**

Since the current is taken to flow in the direction that a positive charge moves, if the current is actually due to the motion of negative charges, then the negative charges flow in the opposite direction, i.e. in the direction of increasing electric potential or against the electric potential gradient. Again, this is where the analogy of fluid flow and electricity breaks down and can be traced to the absence of negative mass.

### 4.3  The role of the battery developed further

Let us consider the effect of placing a positive charge $+Q$ and a negative charge $-Q$ at opposite ends of a conductor. To take a definite example we will assume that the conductor is a piece of metal (see Figure 4.5). Recalling previous work (see Section 3.2), a metal consists of a regular lattice of positively charged atomic cores surrounded by 'free' or 'conduction' electrons which are uniformly distributed throughout the metal. The atomic cores remain more or less fixed in their lattice sites, while the free electrons move more or less freely throughout the lattice and are able to respond readily to any electric forces which may be applied.

According to either the First Law of Electrostatics or Coulomb's law (see Chapter 2) the electric forces due to the charge $+Q$ at one end of the metal and $-Q$ at the ends of the metal cause any 'mobile charge carriers' present to move. In the case of a metal these mobile charge carriers are just the free electrons. The atomic cores (being positively charged) try to move away from the charge $+Q$ and towards the charge $-Q$. However, the atomic cores are very difficult to move and remain fixed at their lattice sites. The free electrons are able to respond to the electric forces and move towards the charge $+Q$ and away from the charge $-Q$. Note that the free electrons move so as to neutralize the effects of the charges $+Q$ and $-Q$, and in a very short time an equilibrium is reached and the electrons come to rest (see Figure 4.6).

We note that the free electrons are now not uniformly distributed throughout the material; there is a surplus of electrons near end A (leaving end A more negative than previously) and a deficit of electrons near end B (leaving end B more positive than previously). Thus, if there has been a change in the distribution of the electrons, this means there has been some period when there has been a movement of electric charge, and therefore, an electric current flows. When the charge ceases

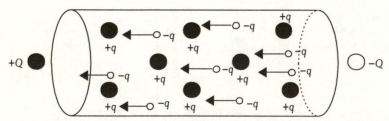

*Figure 4.5*  **The (mobile) charges in the metal move to neutralize the effects of the charges $+Q$ and $-Q$: a transitory current flows**

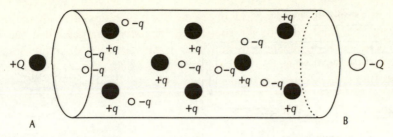

*Figure 4.6* **In equilibrium the charge distribution of the metal has changed**

to move no electric current can be flowing. In this situation the current that flows depends upon time and in this particular case, the current is referred to as a transitory current flow.

In practical circuits it is obviously very useful to have some means of ensuring that a *constant* current flows, i.e. a current which does not depend upon time. Suppose we can somehow move the surplus of electrons from the end A and move them to the end B. What will be the effect of this? These electrons, having been placed near the negative charge $-Q$ will again feel the influence of the repulsive forces due to $-Q$ and so will move away towards the charge $+q$ as before, i.e. a current will again flow until equilibrium is re-established, as described previously. If we could continuously move the surplus negative charge that accumulates at the end A to the end B then the current that flowed would consist of a series of pulses. If we moved the charge from A to B fast enough the current pulses would approximate a continuous current, the approximation getting better the faster we moved the negative charge that migrates from the end A to the end B.

Of course, this would not be a very practical way of achieving a continuous electric current. The role of the battery can be thought of as achieving the above in the following manner: Just before the battery is connected to the metal it is in its normal state where the free electrons are uniformly distributed and moving randomly throughout the metal. Just after the battery has been connected, the free electrons are repelled from the negatively charged terminal, and move away from this terminal. This movement causes nearby electrons further along the piece of metal to be pushed away (since these electrons are, according to the First Law of Electrostatics, repelled by those electrons which have just been pushed away from the negative terminal of the battery). This happens along the length of the metal until, finally, a group of electrons reaches the positive terminal. Note that when the free electrons are repelled from the negative terminal of the battery, a positive charge is left (since if something was originally overall neutral, removing negative charge from it leaves it with a positive charge). The effect of the electrons entering the battery at the positive terminal is to 'push' an equal amount of electric charge out at the negative terminal.

We are now back at the original situation when we just connected the piece of metal to the battery. It is as if, in Figure 4.5, we somehow arranged that the surplus of electrons at the end A were moved continuously back to the end B. In this manner

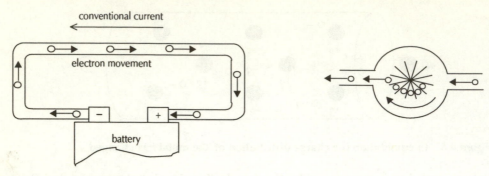

*Figure 4.7* **The battery as an electron pump**

we can think of the battery as an 'electron pump', pushing out negative charge at the negative terminal and accepting the charge at the positive terminal, as shown in Figure 4.7.

In general, we refer to the battery as the *source* of the e.m.f. The battery achieves such a separation of electrical charge by chemical means, i.e. chemical energy is converted into electrical energy. Some further terminology is that we refer to electric current that flows in just one direction as a 'direct current' or DC. Strictly, this terminology tells us nothing about how the current depends on time, but is usually applied to a constant current.

--- COMMENTS

- A positively charged particle would move away from the positive terminal of the battery, towards the negative terminal. Hence, the current in a circuit flows from the positive terminal to the negative terminal of the battery. Conversely, a negatively charged particle moves from the negative terminal to the positive terminal of the battery. The direction of the current is taken to be the direction that a positive charge tends to move, and is the same in both cases.

- Note that if the current leaving the positive terminal is $I$, the current entering the negative terminal is also $I$. This is an example of electric charge conservation, i.e. the total amount of electric charge in any closed system remains constant. If the amount of charge that leaves the positive terminal of the battery in a small time $\Delta t$ is $\Delta Q$, then this amount of charge must enter the battery at the negative terminal in the same time. It then follows from the definition of electric current, $I = \Delta Q/\Delta t$, that the current must remain the same.

- A more precise manner of discussing the problem is to argue that the battery provides a difference in electrical potential, the terminal marked '+' is at a higher electric potential than the terminal marked '−'. Electric current flows from points with a difference in electrical potential, i.e. between points with a potential difference, flowing from regions of high electrical potential to regions of low electrical potential.

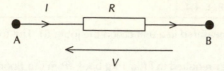

*Figure 4.8*  **The arrow notation relating to potential difference and current flow**

### The arrow notation relating potential difference and current flow

This last comment is particularly important when we discuss Kirchhoff's laws in Chapter 7. With this in mind we introduce an important arrow notation: Figure 4.8 shows the (conventional) electric current flowing from point A through the resistor *R* to the point B. According to the comment, the point A must be at a higher electric potential than the point B, and we show this by placing a line labelled by the amount of the difference in electrical potential between the points, *V* say, with an arrow head, the arrow pointing from the point B to the point A, i.e. the arrow points in the direction of *increasing* electric potential.

> Since the convention depends upon conventional current flow, this notation is independent of whether the current is due to the movement of positive or of negative charges.

### 4.4  Electromotive force and potential difference

It follows from previous discussions that we can think of a charge as moving between two points due to electric forces provided by the battery in terms of the *difference* in the electric potential of the points. Thus, there is some link between 'electric forces' pushing a charge between two points and the 'electric potential' between them. The link is the amount of work done to move the charge between the two points. This is reasonable, since when forces move a particle, work is being done on the particle. Here, electric forces act on charges, forcing the charges to move, i.e. the electric charges have work done on them by the electric forces.

The amount of work done on the charge depends not only on the difference in electrical potential between the points, but also on the size of the charge the electric forces act upon. We quote the work done by the electric forces, *W*, to move a particle of charge *Q* between two points with a potential difference *V* between them as

$$W = Q \times V \tag{4.1}$$

Thus, Equation (4.1) tells us that when electric forces move a charge between two points, the work the forces do in moving the charge depends upon the potential difference between the two points, *V*, and also on the size of the charge, *Q*. Similarly, if we push a particle up a hill against gravity, the amount of work done

■■■■ **WORKED EXAMPLE 4.1**

All forms of energy are measured in a unit called the joule (J). The following examples indicate the size of this unit:

(a) The amount of energy required to lift a 1 kg book from the floor onto a table 1 m above it is about 10 J.

(b) The amount of energy converted in heating one litre of water is about 0.5 MJ.

depends not only upon the height we push the particle through, but also on its mass. If we want to move a particle of mass $m$ through a height $h$, the work that has to be done in moving the particle against the force of gravity is given as

$$W = m \times g \times h$$

where $g$ is the acceleration due to gravity.

The unit of potential difference is therefore the joule per coulomb (J $C^{-1}$). This is more conventionally referred to as the volt (V). Thus

$$1\,V \equiv 1\,J\,C^{-1}$$

■■■■ **DEFINITION**

Definition: We define the volt as the p.d. difference between two points if one joule of work is used to move one coulomb of charge between the points.

Figure 4.9 shows that a charge of one coulomb gains one joule of energy when it moves between two points with a difference in electric potential of one volt.

The term 'electromotive force' is taken to indicate the amount of work done by the battery to move charge *completely* around the circuit, while the term 'potential difference' is taken to refer to the work done to move charge between *two points* in a circuit.

■■■■ **DEFINITION**

Definition: We define the e.m.f. of a source as the total work done in joules to move one coulomb of charge completely around the circuit.

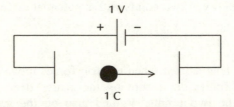

*Figure 4.9*   **A charge of 1 C gains 1 J of energy from the electric forces**

Note the similarities and differences between the definitions of p.d. and e.m.f; potential difference refers to two points in the circuit, while electromotive force refers to the whole circuit.

Before discussing circuits proper, we will spend some more time discussing this definition of potential difference. Figure 4.10 shows a particle of mass $m$ and charge $q$ placed between two plates which are connected across a battery of e.m.f. $E$. In this case we have the p.d. between the two plates as $E$, i.e. $V = E$ (this will be discussed again in Chapter 5). The particle, being charged, is subjected to the electric forces between the plates. It follows that the particle has work done on it by the electric forces, and therefore its kinetic energy increases, and, hence, its speed increases. If the particle is positively charged and is placed near the plate connected to the positive terminal, the force acting on it will be directed towards the plate connected to the negative terminal. The amount of work done on the charge is given simply as

$$W = qV$$

This follows from the definition of the volt; see Equation (4.1).

### ■■■ KINETIC ENERGY

> The kinetic energy of a body is the energy the body possesses by virtue of its motion. If a body is initially at rest we have to apply forces to the body for it to move, i.e. we have to do work on the body. The amount of work we have to do so that a body of mass $m$, initially at rest, moves with speed $v$ is equal to the kinetic energy that the body possesses. This is given as
>
> $$\tfrac{1}{2} m v^2$$

The work done by the electric forces is used to increase the kinetic energy of the particle. If the particle starts off from rest and reaches a final speed $v$, the initial kinetic energy is zero and the final kinetic energy is $\tfrac{1}{2} m v^2$, so the *change* in kinetic energy is $\tfrac{1}{2} m v^2$. Since this change in energy is equal to the amount of work done on the particle we have

$$\tfrac{1}{2} m v^2 = qV$$

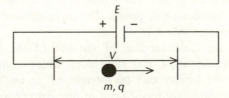

*Figure 4.10*  **A particle of mass $m$ and charge $q$ starts from rest and is accelerated through the p.d. $V$**

━━━ **WORKED EXAMPLE 4.2**

A particle of mass $m = 9.1 \times 10^{-31}$ kg and charge $-1.6 \times 10^{-19}$ C is accelerated from rest through a potential difference of 1000 V. Determine its final speed.
– Making use of the formula (4.1), we have

$$v = \sqrt{\frac{2 \times 1.6 \times 10^{-19} \times 1000}{9.1 \times 10^{-31}}} = 1.88 \times 10^7 \, m\,s^{-1}$$

Note that we are told that the particle is being accelerated, so that the signs of $q$ and $V$ have been taken correctly into account. Also, since we have used SI units throughout, we are assured that the final result is in m s$^{-1}$.

From this important result we can determine the final speed of the particle in terms of the accelerating potential $V$ as

$$v = \sqrt{\frac{2qV}{m}}$$

## 4.5  Summary

Having introduced electric charge and then discussed charge in motion we made use of a simple electric circuit to introduce several important terms in electric circuits. In particular, the role of the battery (source of e.m.f.) was discussed. After this the terms electromotive force and potential difference were contrasted and compared. By means of Equation (4.1), we note that both refer to the amount of work done to move charge between two points. In the case of the e.m.f., this indicates the amount of work done to move charge completely around the circuit, while the p.d. indicates the amount of work done to move charge between two points on the circuit, for example to move charge from one side of a resistor to the other.

The role of resistance was introduced as one factor which determines the amount of current flowing through a circuit.

An important analogy with a fluid moving through a pipe was made. We suggested that electric charge moves between two points because of a difference in electric pressure between the points, i.e. a potential difference between two points causes charge to move between those points. The main similarities between fluid and electric circuits are listed in Table 4.1.

We then developed our understanding of the role of the battery by noting that ultimately it is the application of electric forces which cause electric charges to move. We found that the battery can be thought of as an 'electron pump', with negative charge being continually pushed out of the battery at the negative terminal and entering at the positive terminal. Finally, we introduced a link between the electric forces causing charges to move between two points and the difference in the

*Table 4.1*

| Fluid circuit | Electric circuit |
|---|---|
| ■ differences in the hydrostatic pressure between points causes fluid to move between the points | ■ differences in the electric pressure between points cause charge to move between the points |
| ■ the pump causes fluid to move around the pipes | ■ the battery causes charge to move around the circuit |
| ■ the amount of water that leaves the pump each second is the same as enters the pump | ■ the current that leaves the positive terminal is the same as the current that enters the negative terminal |
| ■ constrictions in the pipe inhibit the fluid flow | ■ resistors in the circuit inhibit the movement of charge, i.e. the current |

electric potential between those points. This enabled us to discuss the similarities and differences between the electromotive force and potential difference.

To proceed further with our discussion of electric circuits, we now have to consider the effect of the resistors in the circuit. We will find that we can link together the p.d. across a resistor and the current flowing through the resistor, and hence through the circuit as a whole. This will be discussed in Chapters 5 and 6.

## ▬ Problems

**4.1**    A particle of mass $1.67 \times 10^{-27}$ kg and charge $1.6 \times 10^{-19}$ C is accelerated from rest through a p.d. of $V$. If the final speed of the particle is $10^6$ m s$^{-1}$ determine the accelerating potential $V$.

**4.2**    A particle of mass $1.67 \times 10^{-27}$ kg and charge $1.6 \times 10^{-19}$ C is accelerated from rest through a p.d. of 1000 V. (a) How much energy does the particle gain? (b) Determine the final speed of the particle. If the particle had an initial speed of $4.38 \times 10^5$ m s$^{-1}$ and is moving in the same direction as the electric force accelerates the particle, determine (c) the energy gained by the particle and (d) the final speed of the particle. *Hint:* Determine the initial kinetic energy. Then use the results of part (a) to determine the final kinetic energy.

**4.3**    A particle of mass $1.67 \times 10^{-27}$ kg and charge $1.6 \times 10^{-19}$ C is moving with a speed of $10^7$ m s$^{-1}$. How much work has to be done to stop the particle? If that work is provided by the application of electric forces, what de-accelerating potential $V$ is required?

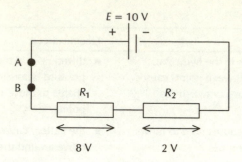

*Figure P4.1*

> *Hint:* Determine the initial kinetic energy of the particle. If the final speed of the particle is zero, the final kinetic energy is zero so that the *change* in kinetic energy is just the original kinetic energy. Equate this to the work that has to be done by the electric forces, making use of Equation (4.1).

**4.4**   Consider the circuit in Figure P4.1. Indicate on the figure the direction of the (conventional) current flow and the voltage arrows across each of the resistors. If the connecting wires are made of metal, indicate the direction of the (conduction) electron flow.

Determine the amounts of energy used to move a charge of 200 mC (a) completely around the circuit, (b) across each of the resistors, and (c) between the points A and B. Comment on your results.

> *Hint:* For part (a) make use of the definition of the volt. For part (c) take the resistance of all connecting wires to be zero so that the p.d. between the points A and B is zero.

o **4.5**   Confirm that $\frac{1}{2}mv^2$ and $Q \times V$ have the same units as energy.
Hint: Make use of the results of Tables 1.1 and 1.3.

# CHAPTER FIVE

# *An introduction to resistance*

Having previously discussed the role of the battery in a simple electric circuit, we now turn to the role of the resistance in the circuit. To help understand the different parts of the circuit, we used the analogy of a fluid flowing through a pipe. This suggested that the amount of current flowing in the circuit depends not only upon the ability of the battery (technically, the source e.m.f.) to move charge around the circuit, but also upon the amount of resistance offered by the circuit to the flow of current.

To proceed further, we now have to consider the effect of the resistors in the circuit in some detail. This will link together the p.d. *across* a resistor and the current flowing *through* the resistor.

After defining the resistance of a component we will be led on to an extremely important empirical relationship discovered by Ohm, which is referred to as Ohm's law. We will make extensive use of this law. We then proceed with a discussion of what resistance depends upon, with some practical aspects of resistors.

## 5.1 The effects of resistance in an electric circuit

We begin with a single resistor, $R$, connected to a battery of e.m.f. $E$ as shown in Figure 5.1. Also shown in the figure are an ammeter measuring the current flowing through the circuit, and therefore the resistor $R$, and a voltmeter measuring the p.d. across the resistor. We do not discuss the principles of operation of either the ammeter or the voltmeter here. We merely note that their operation depend upon some effect of the current flowing through them. They are assumed to be perfect instruments that do not influence the circuit in any manner.

Suppose the ammeter measures the current flowing through the circuit as $I$ and the voltmeter measures the p.d. across the resistor as $V$. Note that since there are no other components in the circuit (the wires connecting the battery to the resistor have a negligible effect on the circuit; this will be justified shortly), then in this case the p.d. measured by the voltmeter, $V$, will be the same as the e.m.f. of the battery, $E$.

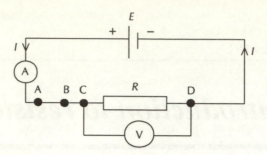

*Figure 5.1*  **A simple circuit consisting of a single resistor connected to a single battery**

This result follows from the definitions of e.m.f. and p.d. in terms of the work done by the electric forces to move a charge completely around the circuit and across the component, respectively. The work done to move a charge $Q$ completely around the circuit is just $QE$ (this is the definition of e.m.f.), while the work done to move the charge across the resistor $R$ is $QV$ (this is the definition of p.d.), since the p.d. across $R$ is $V$, (see Equation (4.1)). If the wire offers no hindrance to the movement of electric charge, then no work is required to move charge along the wire. In particular, consider a section A, B of the wire as shown in Figure 4.5. If no work is done to move the charge across this piece of wire, then the total work done to move the charge completely around the circuit is the same as the work done to move the charge across the resistor. Therefore we have

$$\begin{pmatrix} \text{work done to move} \\ \text{charge across} \\ \text{resistor} \end{pmatrix} = \begin{pmatrix} \text{work done to move} \\ \text{charge completely around} \\ \text{circuit} \end{pmatrix}$$

$$QV = QE$$

Cancelling out the charge $Q$ on both sides of this equation gives the desired result. Alternatively, an argument based on the concept of 'electric pressure' may be used.

It is found that the current flowing through the circuit depends not only on the e.m.f. of the battery, but also on a property of the resistor referred to as its resistance. This property reflects the amount of restriction offered to the flow of current. As suggested previously (see Section 3.2) there are two broad classes of materials: conductors and insulators. The reason for this difference in the behaviour of materials is the presence or absence of mobile charge carriers. Metals, for example, have available many free electrons which are able to contribute to an electric current, while plastics, for example, have only very few free electrons available.

Returning to the discussion, we *define* the resistance of the resistor by the following formula:

$$V = IR \tag{5.1}$$

Thus, to determine the resistance of some component we simply measure the p.d. across it and divide this by the current flowing through it, i.e.

$$R = \frac{V}{I} = \frac{\text{p.d. } across \text{ component}}{\text{current } through \text{ component}} \qquad (5.2)$$

If the p.d. $V$ is measured in volts (V) and the current $I$ is measured in amperes (A) then $R$ will be determined in volts per ampere (V A$^{-1}$). This is more often referred to as the ohm ($\Omega$), the SI unit of resistance:

$$1\,\Omega \equiv 1\,\text{V A}^{-1}$$

━━━ **WORKED EXAMPLE 5.1**

A battery of e.m.f. 5 V is connected across a resistor and a current of 10 mA is measured flowing through it. Determine the resistance.

– According to definitions (4.1) or (4.2) the resistance of the resistor is given by

$$R = \frac{V}{I} = \frac{5\,\text{V}}{10 \times 10^{-3}\,\text{A}} = 500\,\Omega$$

We can now offer a different argument as to why the connecting wires do not influence the flow of current in a circuit. The wires connecting the resistor to the battery are taken to be 'perfect' and to have zero resistance. Let us consider again the section A,B of the connecting wire. The same current $I$ flows through this section of the wire as through the resistor. Let us determine the p.d. across this piece of wire, call it $V_{AB}$. We have, from the definition (5.1),

$$V_{AB} = IR_{AB}$$

where the resistance of the section A,B is $R_{AB}$. But, we have said that the resistance of the wire is zero, so that $R_{AB}$ is equal to zero. This gives $V_{AB}$ equal to zero, and so no work is done to move charge across this section (from Equation (4.1)). There are two points to note here:

■ In practice the connecting wires will have a very small, but non-zero resistance. According to the above argument the wires *will*, therefore, affect a circuit. This effect will, however, usually be very small. We will generally assume that the resistance of connecting wires will always be much smaller in practice than the resistance of the other components of interest, i.e. the connecting wires will have only a negligible effect and the connecting wires *do not* affect a circuit.

■ If the potential difference between the two points A and B is zero, then the points are said to be at the *same* potential.

### Resistance and Ohm's law

We note that Equation (5.2) *defines* the resistance of a component. It tells us how we experimentally determine the resistance, i.e. we measure the p.d. across and the current flowing through the component, then divide the p.d. by this current. It is important to be aware that Equation (5.2) does not imply that the resistance is a constant and independent of the p.d. across it. Thus, in Worked Example 5.1 there is nothing to suggest that if a battery of e.m.f. 10 V is connected across the resistor then a current of 20 mA will flow through the resistor.

If we *knew* that the measured value of the resistance was a constant, then we could predict that a current of 20 mA would flow through the resistor when a battery of e.m.f. 10 V was connected across it.

Ohm discovered that for many materials of interest the following is true:

---

'For a wide class of materials of interest under constant physical conditions, $R$ is a constant'

$R = \text{constant} \Rightarrow I \propto V$

---

This result is known as Ohm's law. Note the term *constant physical conditions* – this will be discussed later. We point out that Ohm's law is more an extremely useful and widely used experimental fact than a 'law' in the conventional sense.

For a component which obeys Ohm's law, referred to as *Ohmic*, if we increase the p.d. $V$ across it, proportionally more current will flow, i.e. if we increase the 'electric pressure' across the component, proportionally more electric charge will flow each second through the component. Figure 5.2 shows how the current, $I$, flowing through a component depends on the p.d. across it for an Ohmic component. Such a figure is referred to as the $I$–$V$ characteristics of the component and is obtained from experiment. The $I$–$V$ characteristics just tell us how the current flowing through the component varies with the potential difference across it. As can be inferred from Figure 5.2, once we have measured or determined the current $I$ for

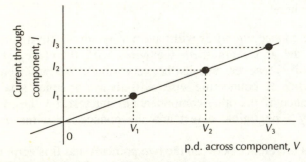

*Figure 5.2*  **$I$–$V$ characteristics for an Ohmic component**

some particular p.d. *V*, then we know the value of the current for any other value of the p.d. The graph of *I* versus *V* is referred to as linear, and shows that as the p.d. is increased across the resistor, the current flowing through it increases proportionally.

Note that the graph passes through the origin. This tells us that if the p.d. across the component is zero then no current flows. This is reasonable and emphasizes previous comments that it is a *difference* in electrical potential that causes electric charge to move between the two points, i.e. for a current to flow.

It is also of interest that the graph is symmetrical about the origin. What does this tell us? If the p.d. across the component is negative this means, according to Figure 5.2, that the current is negative. Recalling Section 3.1, this means that the current is flowing in the opposite direction to which we assumed. Thus, the result tells us that if we reverse the battery connections, the current flows in the opposite direction, which is just common sense. The result tells us a bit more than this, since it also tells us that the current is still proportional to the applied p.d. We will soon see that there are components which behave quite differently according to how the battery is connected across them. Further details on graphs are given in the Appendix.

We will make use of Ohm's law in two definite circumstances once the resistance *R* is given: either

- we know the p.d. across a component and we wish to determine the current flowing through it, given as

$$I = \frac{V}{R}$$

or,
- we know the current flowing through the component and we wish to determine the p.d. across it, given as

$$V = IR$$

The relationship between p.d., current and resistance is conveniently remembered by making use of the so-called 'Ohm's law triangle'; see Figure 5.3. This triangle is to be used in the following manner. Suppose that the p.d. across a component and the resistance are known, and we wish to determine the current flowing through the component. Simply cover the current *I* in the triangle and we note that we are left with *V*/*R*, see Figure 5.4.

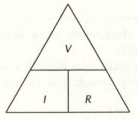

*Figure 5.3*   **Ohm's law triangle**

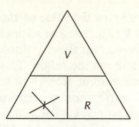

*Figure 5.4* **A simple use of the Ohm's law triangle to show** $I = V/R$

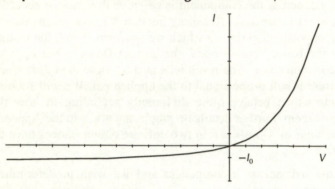

*Figure 5.5* *I–V* **characteristics for a diode**

### Non-Ohmic components

We complete our discussion of the effects of resistance in a circuit by considering a particular component which does not obey Ohm's law. Such a component is referred to as non-Ohmic and its *I–V* characteristics, in this example, take the form depicted in Figure 5.5.

Notice that for *V* negative the current flowing through the diode is practically independent of the p.d. across it, while for *V* positive the current increases more rapidly than linearly. This graph is referred to as non-linear. Again, that the curve passes through the origin means that if we apply no p.d. no current flows, which is reasonable. That the curve is no longer symmetrical about the origin means how the diode operates depends upon how the battery is connected to it. Contrast this behaviour with that for an Ohmic material shown in Figure 5.2. This has important consequences and is discussed further in Chapter 14.

■■■■ **WORKED EXAMPLE 5.2**

A particular diode has the *I–V* characteristic curve shown in Figure 5.6. Determine the resistance at the points A, B, C and D.
– Let us consider point A in some detail. From Figure 5.6 we see that when the p.d. across the diode is 0.1 V the current flowing through it is 130 nA. Thus, from Equation (5.2), the resistance at this particular p.d. is

$$R = \frac{V}{I} = \frac{0.1 \text{ V}}{130 \times 10^{-9} \text{ A}} = 0.769 \ \Omega$$

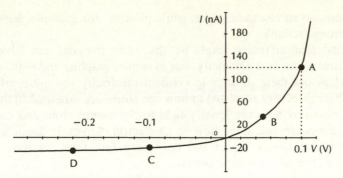

*Figure 5.6* **I–V characteristics for a diode**

The values of the resistance for all points indicated in Figure 5.6 are tabulated in Table 5.1.

*Table 5.1*

| Point | p.d. (V) | I (nA) | R (MΩ) |
|-------|----------|--------|--------|
| A | 0.1 | 130 | 0.769 |
| B | 0.05 | 35 | 1.429 |
| C | −0.1 | −18 | 5.555 |
| D | −0.2 | −20 | 10.000 |

An important feature of Worked Example 5.2 is that the resistance of the diode depends upon the p.d. across it. This means we cannot make use of Ohm's law to determine the current flowing through the diode since the resistance now depends upon the applied p.d. across it. This is in marked contrast with the behaviour of a component for which Ohm's law is valid. In such a case, once the resistance has been determined from one p.d. and current measurement, it is known that for all applied p.d.'s, and if we change the p.d. across the component we can determine the current flowing through it.

## 5.2 What resistance depends upon

We now wish to discuss the factors upon which the resistance depends. Experimentally, we find that the main factors are:

### The type of material

This is reasonable, since we have noted previously that the ability of a material to conduct an electric current depends partly on the presence or absence of mobile charge carriers. Metals, for example, have available many free electrons which are able to contribute to an electric current, while plastics, for example, have only very few free electrons available.

We also find that different forms of the same material can have markedly different abilities to conduct electricity. For example, graphite and diamond are both forms of carbon but their abilities to conduct electricity are quite different. We merely note that this ability is related to how the atoms are arranged in the lattice. In graphite, the electrons are only loosely held to the parent atom and can easily be 'shaken' free and contribute to the flow of an electric current. In diamond, however, all the electrons are tightly bound and only very few free electrons are able to contribute to the flow of an electric current.

Given that a material has free electrons and so it can conduct an electric current, what is the mechanism by which the material offers resistance? Our simple model of a conductor introduced in Section 3.2 again offers a suggestion. Suppose a battery is connected across the ends of a piece of conductor. As explained previously this will cause the free electrons in the material to flow. As the electrons move through the material they occasionally collide with the atomic cores fixed at the lattice sites. These collisions cause the electrons to lose some of their energy and, therefore, their movement is restricted (see Figure 5.7). The more the electrons collide with the atomic cores the greater will be the restriction to the flow of the electrons. The amount of the restriction, therefore, depends not only upon the number of free electrons available, but also on how the atomic cores are arranged in the lattice, which depends upon the material.

### The length of the material through which the current flows

It has been found experimentally that as the length of a conductor is increased, the more opposition it presents to the flow of an electric current, i.e. the resistance increases. Recalling the analogy with fluid flowing in a pipe and electric current flow this is easy to understand – we know from experience that it is easier to push a given quantity of fluid through a short pipe than through a long one.

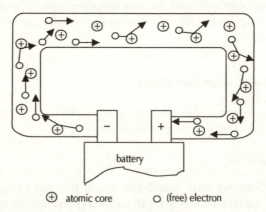

$\oplus$ atomic core    $\circ$ (free) electron

*Figure 5.7* **The current is restricted as it tries to flow**

### The thickness of the material through which the current flows

Similarly it has been found that the resistance depends upon the cross-sectional area of the conductor. In particular, experiments have shown that as the cross-sectional area of the conductor increases the resistance decreases. Again, the analogy with fluid flow is helpful – we know from experience that it is easier to pass a given quantity of fluid through a wide pipe than through a narrow capillary.

### The temperature of the material

When we quoted Ohm's law we were careful to include the phrase 'under constant physical conditions'. The most important physical condition that resistance depends upon is temperature. Thinking of a metal as consisting of a lattice of atomic cores surrounded by free electrons, we note that as the temperature of the metal increases the atomic cores vibrate more violently about their lattice sites and, therefore, have more chance of knocking an electron out of its path. Since an electric current is merely the flow of electric charge, any process that impedes the flow of charge will affect the ability of the metal to conduct electricity, so that we might expect the resistance to depend upon temperature. In particular, we expect the resistance to increase with temperature.

## The resistivity

As discussed previously, the analogy of fluid flowing through a pipe suggests that the resistance of the material should depend on the length of the material in such a manner that as the length gets larger then the resistance increases. Similarly, the analogy suggests that as the cross-sectional area of the material increases, the resistance becomes smaller. To determine the exact dependence of this relationship we have to appeal to experiment. Let us consider a component made of some given material of various lengths, $l$, and of various cross-sectional areas, $A$. Then let us place a source of fixed e.m.f. across the component and measure the current flowing through and the p.d. across it, say $I$ and $V$, respectively, as shown in Figure 5.8 (in the figure do not confuse the symbol $A$ representing cross-sectional area and the A representing the ammeter; as noted previously, how a symbol is to be interpreted should be obvious from the context).

Careful experiments show that if the cross-sectional area of the material is kept fixed, then the resistance $R$ increases linearly with the length, $l$, of the material:

$$R \propto l$$

This means that if a current $I$ flows through the component due to a p.d. $V$, then if its length is doubled, only half as much current flows as before. If half the current now flows while the p.d. has been kept constant, then, according to Equation (5.2), the resistance has doubled.

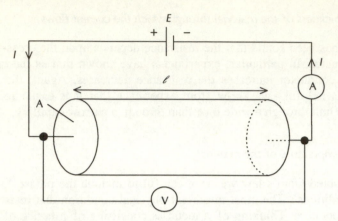

*Figure 5.8*    **Introducing resistivity**

Similarly, if we keep the length of the material fixed, and measure how the current varies as we change the cross-sectional area, we find that the resistance is inversely proportional to the cross-sectional area:

$$R \propto \frac{1}{A}$$

Thus, if a current $I$ flows when the p.d. is $V$, twice the current flows when the area is increased by a factor of two, i.e. the resistance has halved.

Putting these two results together we find the resistance depends on the length and cross-sectional area in the following manner:

$$R \propto \frac{l}{A}$$

How do we take account that the resistance also depends on the type of material? We simply introduce a quantity known as the resistivity, $\rho$, as a constant which turns the above proportionality into an equality, and write

$$R = \rho \frac{l}{A} \qquad (5.3)$$

The resistivity is specific to the material and contains all the information about the number of electrons each atom of the material makes available for conduction and how the atoms are packed together. When we (loosely) say that, for example, copper offers less resistance to the flow of current than does, say, plastic, what is really meant is that the resistivity of copper is less than that of plastic. The resistivity may be thought of as a kind of 'specific resistance' of a material, i.e. the resistance of a material where the geometric factors of length and cross-sectional area have been accounted for.

Note that normally we only quote the resistivity for Ohmic materials. The reason for this is straightforward, since, otherwise, for a non-Ohmic material, the resistivity would depend on the p.d., and it would be necessary to quote the value of the resistivity for each p.d., which would be cumbersome.

## The units of resistivity

We can write Equation (5.3) in terms of units as follows:

$$R \ [\Omega] = \rho \ [?] \ \frac{l \ [\text{m}]}{A \ [\text{m}^2]}$$

since the SI units of resistance, length and cross-sectional area are the ohm ($\Omega$), metre (m) and square metre (m$^2$), respectively. From this equation we deduce that for the units to balance we must have the units of resistivity to be the ohm-meter ($\Omega$ m).

Note that if we take a unit cube of material, i.e. $l = 1$ m, $A = 1$ m$^2$, then the resistance is numerically equal to the resistivity. Table 5.2 indicates the ranges of values of the resistivity for various materials (in all the examples the temperature is taken to be 20°C; see later). Of the materials listed in Table 5.2, copper and nichrome are examples of metals, germanium and graphite are examples of semiconductors, while diamond is an example of an insulator.

Worked Example 5.3 is a straightforward numerical exercise.

---

■■■■ **COMMENTS**

---

- From Table 5.2, the resistivity of copper, $\rho_{Cu} = 1.72 \times 10^{-8} \ \Omega$ m, is some 65 times smaller than the resistivity of nichrome, $\rho_{Ni} = 112 \times 10^{-8} \ \Omega$ m, i.e.

$$\frac{\rho_{Cu}}{\rho_{Ni}} = \frac{1}{65}$$

This implies that copper is a better conductor than nichrome for wires of the same length and cross-sectional area. This follows since for given length, $l$, and cross-sectional area $A$, the resistances of wires made from copper and nichrome are given as, respectively

$$R_{Cu} = \rho_{Cu} \frac{l}{A}, \qquad R_{Ni} = \rho_{Ni} \frac{l}{A}$$

so that we have

$$\frac{R_{Cu}}{R_{Ni}} = \frac{1}{65} \Rightarrow R_{Cu} = \frac{1}{65} R_{Ni}$$

- According to Table 5.2 the resistivity of germanium (at 20°C) is approximately $10^{-3} \ \Omega$ m. Replacing just one germanium atom out of a million with a phosphorus atom (a process known as 'doping'), the resistivity is decreased by a factor of 1000. This shows the extreme sensitivity of the resistivity of germanium to impurities.

■ Graphite and diamond are both forms of carbon, yet the resistivities of these materials differ by a factor of approximately $10^{16}$. As mentioned previously, this can be related to the different manner in which the carbon atoms are arranged in graphite and in diamond. The range of values quoted for graphite is due to the sensitivity of the resistivity to the presence of small amounts of impurities, while that quoted for diamond is related to the experimental difficulty of measuring such a large resistivity.

Table 5.2

| Material | $\rho$ ($\Omega$ m) |
|----------|---------------------|
| copper | $1.72 \times 10^{-8}$ |
| nichrome | $112 \times 10^{-8}$ |
| germanium | $\sim 10^{-3}$ |
| graphite | $\sim (3-60) \times 10^{-6}$ |
| diamond | $\sim 10^{10}-10^{11}$ |

**WORKED EXAMPLE 5.3**

Determine the resistance of a sample of copper of length $l = 100$ m and of uniform radius $r = 0.4$ mm. What is the resistance if the radius changes to $r = 0.5$ mm?

− The length and resistivity of the material are given; we only have to determine the area and make use of Equation (4.3). We have $A = \pi r^2 = 0.5$ mm$^2 = 0.5 \times 10^{-6}$ m$^2$; hence

$$R = \rho \frac{l}{A} = 1.72 \times 10^{-8} \times \frac{100}{0.5 \times 10^{-6}} = 3.44\Omega$$

To determine the resistance when the radius changes from 0.4 to 0.5 mm we make use of the fact that the resistivity depends inversely on the area, which in turn depends upon the square of the radius, and is therefore given as

$$\frac{(0.4)^2}{(0.5)^2} \times 3.44\Omega = 2.20\Omega$$

In the second part of the example it was straightforward to make use of the fact that the resistance varies as the inverse of the area. The alternative is to determine the area and make use of Equation (5.3) directly.

### Temperature dependence

As noted previously, we expect the temperature to affect the ability of a metal to conduct an electric current, i.e. we expect the resistance to depend upon temperature, since resistance measures this ability. The argument also suggested that as the temperature of a metal increases, the ability of the metal to conduct is inhibited. Thus, we not only expect the resistance of a metal to depend upon temperature, we also expect the resistance to increase with temperature.

Since resistivity is more closely related to the material itself, it is conventional to first discuss the temperature dependence of the resistivity rather than the resistance. That the resistivity of a conductor depends upon temperature is allowed for if we give a rule on how it changes with temperature. We find the rule describing the change of resistivity with temperature is very simple and is given as

$$\rho(T) = \rho(20°C) \times (1 + \alpha \times (T - 20))) \tag{5.4}$$

In this formula $\rho(T)$ means 'the resistivity that depends on the temperature' and $\rho(20°C)$ is the value of the resistivity at the particular temperature of 20°C, this choice being quite arbitrary. The quantity $\alpha$ (the Greek symbol *alpha*) is called the temperature coefficient of the resistance (even though we are discussing the temperature dependence of the resistivity) and is quoted in standard tables for many materials.

Given that we now have a rule for how resistivity changes with temperature, we can determine the manner in which resistance depends upon temperature. This follows since resistance and resistivity are related by the result (5.3), i.e. $R = \rho(l/A)$. If $l$ and $A$ are taken to be constant over the temperature change we have

$$\rho(T) \frac{l}{A} = \rho(20°C) \frac{l}{A} \times (1 + \alpha \times (T - 20))$$

but the quantity $\rho(l/A)$ is just the resistance of the sample, so that $\rho(T) \, l/A$ is the resistance of the resistor at the temperature $T$, i.e. $R(T)$, while $\rho(20°C) \, l/A$ is the resistance at 20°C, i.e. $R(20°C)$. We therefore have that

$$R(T) = R(20°C) \times (1 + \alpha \times (T - 20)) \tag{5.5}$$

i.e. the resistance has the same temperature dependence as the resistivity.

### The units of the temperature coefficient

From the result (5.4) we must have that the quantity

$$\alpha \times (T - 20)$$

does not have any units. This follows since the quantity $\alpha \times (T - 20)$ is added to a pure number in Equations (5.4) and (5.5). This must mean that $\alpha$ has the units of inverse temperature. Strictly, the SI unit of temperature is the kelvin (K), although in practice the degree Celsius (°C) is still often used for temperature. It therefore follows from the above that an appropriate unit of the temperature coefficient of

*Table 5.3*

| Material | α (°C)$^{-1}$ |
| --- | --- |
| copper | 0.004 28 |
| aluminium | 0.004 35 |
| iron | 0.006 25 |
| manganin | 0.000 01 |
| carbon | −0.000 5 |

resistance is the per degree Celsius, (°C)$^{-1}$. We note that a *change* in temperature of one kelvin is equal to a change in temperature of one degree Celsius.

Typical values of α for some common conductors are given in Table 5.3. We note that the metals copper, aluminium and iron all have positive temperature coefficients. Thus, for these materials, their resistance increases with temperature as expected. The alloy manganin has a very low temperature coefficient, which implies that the resistance of manganin only depends very weakly upon temperature. One important use for such a property is in the manufacture of high precision resistors. Finally, carbon has a negative temperature coefficient. Thus, the resistance of carbon *decreases* with increasing temperature. This is related to the conduction process in semiconductors and will be discussed briefly in Chapter 14. This negative resistance–temperature dependence is put to good use in 'thermistors' – temperature-sensitive devices which compensate for the rise in resistance of other circuit components. We point out that care should be taken when using Equations (5.4) and (5.5) for the temperature dependence of semiconductors since the results are only approximately valid for those materials.

Worked examples 5.4 and 5.5 provide straightforward numerical exercises.

███ **WORKED EXAMPLE 5.4**

Given $R(20°C) = 14\ \Omega$ determine the resistance at 120°C if the temperature coefficient of the resistance α = 0.005 (°C)$^{-1}$.
– Making direct use of Equation (5.5), we find that the resistance at the temperature $T$ is given as

$$R(T) = 14 \times (1 + 0.005 \times (120 - 20)) = 14 \times (1 + 0.5) = 21\ \Omega$$

Given the temperature, $T$, it is straightforward to determine the resistance $R$, i.e. $R(T)$. An alternative problem is given the resistance, determine the appropriate temperature.

The Appendix discusses some aspects of solving equations.

■■■■■ **WORKED EXAMPLE 5.5**

Given $R(20°C) = 14\ \Omega$ and the resistance at some unknown temperature $16.8\ \Omega$, determine the temperature if the temperature coefficient of the resistance $\alpha = 0.025$ (°C) $^{-1}$.
− We have $R(T) = 16.8\ \Omega$ and $R(20°C) = 14\ \Omega$ and we wish to determine the temperature $T$, i.e. that value of $T$ for which

$$16.8 = 14 \times (1 + 0.025 \times (T - 20))$$

It is straightforward either to manipulate the above equation for the unknown temperature $T$ or to rearrange (5.5) to yield

$$T = \frac{1}{\alpha} \frac{R(T) - R(20°C)}{R(20°C)} + 20$$

In both cases we determine $T$ as 28°C.

## 5.3 Some practical aspects of resistors

There are two main types of resistor: the fixed and the variable value resistor. As their names suggest, fixed value resistors are manufactured to have a particular value, while variable value resistors are designed so that their resistance values can be changed by some mechanism. Since we have no need to discuss variable value resistors we restrict our discussion to fixed value resistors.

Fixed value resistors are available with a large set of agreed values (referred to as *preferred values*) and are constructed using various methods and values. There are four main types of fixed value resistor:

- carbon-composition resistor
- metal-film resistor
- wire-wound resistor
- integrated resistors

We discuss each of these in turn:

*Carbon-composition resistor.* This type of resistor is made with a mixture of finely ground carbon that conducts electricity and an insulating material such as clay. The ratio of carbon to the insulating filler determines the resistance.

*Metal-film resistor.* This type of resistor is the most widely used in electronics. It consists of a very thin layer of metal or carbon coated on a ceramic cylinder. A spiral is then cut into the film to produce a very long path. The resistance is determined by the length of the path; the longer the path the larger the resistance.

*Wire-wound resistor.* This type of resistor is made by winding a fine insulating wire around an insulating former. These resistors are used in situations where the circuit has to cope with large currents since they have relatively high power ratings (see Section 6.5 for a discussion of power).

*Integrated resistors.* This type of resistor is made up of strips of semiconductor material in a monolithic circuit. The resistance depends upon the cross-sectional area and the length of the strip. This type of resistor has the advantage of being very small, and is used in compact assemblies.

## Resistor colour codes

Fixed value resistors generally use an international coding system which indicates both the value in ohms and the resistor tolerance. Resistors with value tolerances of 5%, 10% and 20% are colour coded with four bands. This colour code band system is shown in Figure 5.9, and the colour code band scheme is given in Table 5.4. The colour code is read as follows:

■ The band closest to the end of the resistor is the first band and represents the first digit of the resistance value. This end of the resistor is referred to as the banded end. If it is not clear which this band is, start from the end that does *not* begin with a gold or silver band.
■ The second band is the second digit of the resistance value.
■ The third band is the number of zeros which must be appended to the result, or the multiplier.
■ The fourth band gives the tolerance.

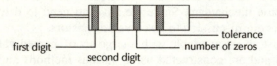

*Figure 5.9* **Colour code bands on a resistor**

*Table 5.4*

|  | Digit | Colour |
|---|---|---|
|  | 0 | black |
|  | 1 | brown |
| Resistance value given by the | 2 | red |
| first three bands: | 3 | orange |
| First band – first digit | 4 | yellow |
| Second band – second digit | 5 | green |
| Third band – number of zeros | 6 | blue |
|  | 7 | violet |
|  | 8 | grey |
|  | 9 | white |
|  | 5% | gold |
| Fourth band – tolerance | 10% | silver |
|  | 20% | no band |

The first two bands of the resistor represent the digits, while the third band represents the number of zeros, and the fourth band represents the tolerance. The first three bands determine the colour coded or nominal value of the resistance. Some real resistors are shown in Figure 5.10.

### Tolerance

Exact values cannot be guaranteed with mass-production techniques, so manufacturers agree to quote the tolerance. What does the tolerance tell us? Although the manufacturer cannot guarantee that the resistor will have the exact value that the colour code indicates, they guarantee that the resistor will be between a certain range of values. This range is determined by the quoted tolerance. In well designed electric circuits the exact values of resistors are not a problem. Thus, in Worked Example 5.6, the fourth band of the resistor was gold indicating that the tolerance is 5%. For this resistor the actual resistance is within ±5% of the colour coded value, i.e. within

$$\frac{5}{100} \, 4700 \, \Omega = 235 \, \Omega$$

of the quoted value. For the resistor under discussion, this means that although the

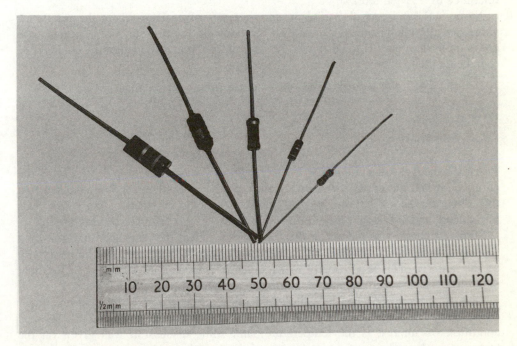

*Figure 5.10*  **Some examples of resistors**

━━━━ **WORKED EXAMPLE 5.6**

A resistor has the colour bands

> yellow, violet, red, gold

Determine the colour coded value of the resistance and the tolerance.
— The first two bands represent the digits: yellow = 4, violet = 7 (i.e. 47). The third band represents the number of zeros: red = 2 (i.e. 00). These zeros are then added to the number to give a nominal resistance of 4700 Ω. The fourth band is the tolerance: gold = 5%. The resistor would be referred to as 4700 Ω (or more usually as 4.7 kΩ) with tolerance 5%.

━━━━━━━━━━━━━━━━━━━━━━━━━━━━━━━━━━━━━━━━━━━━━━━━━

manufacturer cannot claim the resistance is exactly 4700 Ω, they will guarantee that the resistance has a particular value in the range 4465–4935 Ω.

### Other schemes

We note that certain precision resistors with tolerances of 1% or 2% are colour coded with five bands. Beginning at the banded end, the first three bands represent the first three digits, the fourth band is the multiplier, and the fifth band indicates the tolerance. Table 5.4 still applies, although now the colour brown indicates a tolerance of 1% and red 2%.

Numerical labels are also used on certain types of resistor where the resistance value and the tolerance are stamped on the body of the resistor. For example, one common system uses R to designate the decimal point and the letters F, G, J, K, M to indicate the tolerance as follows:

$$F = \pm 1\%, \quad G = \pm 2\%, \quad J = \pm 5\%, \quad K = \pm 10\%, \quad M = \pm 20\%$$

## 5.4 Summary

The resistance, $R$, of a component was defined as the ratio of the p.d., $V$, across the component and the current, $I$, flowing through the component. It was stressed that this definition does not imply that the resistance is constant. Ohm's law was introduced, and although it is not universally valid, it was found to be very useful.

Some emphasis was placed on the factors that affect the resistance offered by a particular material to the flow of an electric current. This introduced the resistivity of the material. This is a property which takes into account the complex factors that determine the ability of a material to conduct an electric current.

One of the provisos for the validity of Ohm's law was that the material under consideration is under constant physical conditions. The most important physical condition of interest is the temperature, and some discussion was given on the temperature dependence of the resistance of a material.

Finally, some practical aspects of resistors were discussed, in particular the types of resistors commonly encountered, together with the meaning of the colour code scheme used to label resistors, and finally the significance of the term tolerance.

## ▬▬ Problems

In the following problems, take the respective resistivities of copper, silver and nichrome at 20°C to be

$$\rho_{Cu}(20°C) = 1.72 \times 10^{-8} \, \Omega\text{m}, \quad \rho_{Ag}(20°C) = 1.65 \times 10^{-8} \, \Omega\text{m},$$
$$\rho_{Ni}(20°C) = 112.0 \times 10^{-8} \, \Omega\text{m}$$

**5.1** In a particular experiment the current flowing through a resistor is recorded as the p.d. across it is varied. The results are as follows:

| $I$ (mA) | 2.0 | 4.0 | 6.0 | 8.0 | 10.0 | 12.0 | 14.0 | 16.0 | 18.0 | 20.0 |
|---|---|---|---|---|---|---|---|---|---|---|
| $V$ (V) | 0.5 | 1.0 | 1.5 | 2.0 | 2.5 | 3.0 | 3.5 | 4.0 | 4.5 | 5.0 |

Plot the current against the p.d. and determine the resistance from the graph.

**5.2** What is the resistance of a piece of copper wire of length 100 m and of uniform radius 1 mm? If the radius of the wire is doubled, *without detailed calculation*, determine the new resistance.
*Hint:* If the length is kept constant, $R \propto \ell/A$, where $A$ is the cross-sectional area.

**5.3** For a given length, $\ell$, and cross-sectional area, $A$, of copper and silver and potential difference between the ends, which material will allow the most current to flow? For the material which offers the least resistance for given values of $\ell$ and $A$, what current will flow through a 100 m length of material of uniform cross-sectional area 1 mm² if a 10 V battery is connected across the ends?

**5.4** The resistance of 100 m of copper wire is 2 Ω. Determine the cross-sectional area of the wire.

**5.5** A piece of copper wire has cross-sectional area 1 mm². What length is needed if the resistance of the wire is 0.5 Ω? If the copper wire is replaced by nichrome wire, what length of the same cross-sectional area would be required to give the same resistance?

**5.6** Given a piece of copper wire of length 100 m and cross-sectional area 2 cm², and a piece of nichrome of length 50 m, what cross-sectional area should the nichrome wire have such that it has the same resistance as the copper wire?

**5.7** A battery of e.m.f. 10 V is connected to a 2 cm length of material of cross-sectional area 0.1 cm². If a current of 2 A flows through the circuit, what is the resistivity of the material?

**5.8** A piece of material has resistance 3 Ω. Given that the material is 20 mm long and has a uniform radius of 2 mm, determine the resistivity of the material.

**5.9**   We have noted that if we take a cube of material with $\ell = 1$ m, $A = 1$ m², then the resistance is numerically equal to the resistivity, $R = \rho$. What would be the resistance if, instead, we took $\ell = 1$ cm, and $A = 1$ cm²?

**5.10**  Given that the temperature coefficient of resistance of copper is $\alpha = 0.00428$ (°C)$^{-1}$ determine the resistivity of copper at the following temperatures: 0°C, 10°C and 50°C.

**5.11**  To what temperature does copper have to be heated or cooled such that its resistivity becomes

$2.46 \times 10^{-8}\ \Omega\text{m}, \quad 1.00 \times 10^{-8}\ \Omega\text{m}, \quad 0.80 \times 10^{-8}\ \Omega\text{m}?$

You are given the temperature coefficient of copper as $\alpha = 0.00428$ (°C)$^{-1}$.

**5.12**  A piece of material has resistance 2 Ω when its temperature is 20°C and a resistance of 5 Ω when its temperature is 50°C. Determine its temperature coefficient of resistance. To what temperature does the wire have to be heated or cooled for a resistance of 4 Ω?

**5.13**  Complete Table P5.1.

*Table P5.1*

| 1st band | 2nd band | 3rd band | 4th band | Nominal resistance | Tolerance | Resistance range |
|---|---|---|---|---|---|---|
| brown | black | brown | gold | | | |
| red | red | green | silver | | | |
| green | blue | green | gold | | | |
| | | | | 27 kΩ | 10% | |
| | | | silver | 100 Ω | | 90 Ω, 110 Ω |
| | | | | 5.6 MΩ | 20% | |

**5.14**  Determine the colour codes for the following resistors:
(a) 150 Ω ± 10%,    (b) 22 kΩ ± 5%,    (c) 330 kΩ ± 20%,    (d) 240 kΩ ± 5%,
(e) 10 Ω ± 10%.

**5.15**  Consider two resistors with quoted values of resistance 20 Ω and 30 Ω. The resistors are connected in series. What current would be drawn from a 5 V battery connected across the resistors? If the manufacturing tolerance for both resistors is 10%, determine the maximum and minimum current which could flow in the circuit.

*Hint:* Consider the two extremes and determine the resistance of the series arrangement if both resistors happened to be at the low end of their tolerance ranges and then if both resistors happened to be at the top of their tolerance ranges.

# CHAPTER SIX

# *Resistance in an electric circuit*

In Chapter 5 we introduced the resistance, $R$, of a component which we defined as the ratio of the p.d., $V$, across the component and the current, $I$, flowing through the component. It was stressed that this definition does not imply that the resistance is constant. Ohm's law was introduced, which states that for a wide class of materials under constant physical conditions, the resistance is constant, i.e. it does not depend on the p.d. across the material. Although not of universal validity, this 'law' is found to be extremely useful in practice.

In this chapter we consider simple resistor networks connected to a single source of e.m.f. and discuss a general approach to determine the currents flowing through and the p.d. across each resistor of the network. A basic assumption made here is that the resistance of each resistor is, indeed, constant.

We begin by discussing the rules for determining the effective resistance of resistors which are placed either in series or parallel. Once these rules have been understood and can be applied, we present a general method which will enable us to solve a great many practical circuits.

Finally, it will then be convenient to discuss the power dissipated by a resistor. We will see that when an electric current flows in a circuit we might expect that electrical energy will be dissipated by the resistor in the form of heat. The rate at which heat is generated in a resistor carrying a current will be discussed.

## 6.1 Introduction

We have discussed resistance and the factors on which it depends, and we now wish to consider the role of resistors in electric circuits. In particular, we discuss how to determine the currents flowing through and the potential differences across each resistor in simple resistor network circuits, such as that shown in Figure 6.1. Such circuits consist of a single source of e.m.f. connected to several resistors.

We begin with the observation that, given a single resistor of resistance $R$ connected to a single source of e.m.f., it is straightforward to determine the current

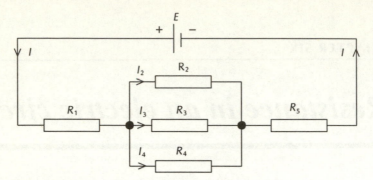

*Figure 6.1*  **A typical resistor network problem of interest**

flowing. Referring to Figure 5.1, and recalling the ensuing discussion, the p.d. across
the resistor is the same as the e.m.f. of the battery, i.e. the p.d. across the resistor is
$V = E$. Therefore the current flowing through the resistor is obtained from Equation
(5.1) as

$$I = \frac{E}{R} \tag{6.1}$$

All such resistor networks of interest to us can be made up of just two special
arrangements of resistor networks:

>   *series* and *parallel*

We will now discuss each of these networks in turn.

### 6.2  Series resistor network

Two resistors $R_1$ and $R_2$ are said to be connected in series if they are connected such
that the same current $I$ flows through each one, as shown in Figure 6.2.

Consider the following circuit which consists of two resistors $R_1$ and $R_2$
connected in series to a battery of e.m.f. $E$; see Figure 6.3. Note that in the figure we
have made use of the arrow notation discussed in Section 4.3. Note that in such an
arrangement, the current flowing through each resistor is the same but the p.d. across
each resistor is different. That the same current flows through each resistor in such a
combination can be made reasonable by considering the analogy of fluid flowing
through two pipes connected in series, as shown in Figure 6.4.

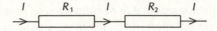

*Figure 6.2*  **Two resistors connected in series**

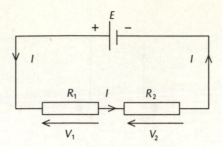

**Figure 6.3    Two resistors connected in series to a battery**

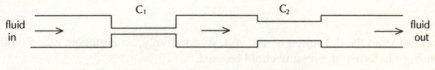

The total flow of fluid depends upon both constrictions:

rate of fluid in = rate of fluid out

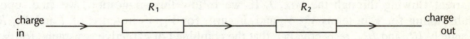

In the electric circuit replace 'fluid' by 'charge' and recall that the current is the rate of flow charge:

the rate at which charge enters = rate at which charge leaves

**Figure 6.4    The analogy between fluid flow and electric current developed further**

We now proceed to derive the required result. The following derivation may be omitted, but the result (6.2) is important.

○    ### Derivation of the result

The p.d. across each resistor is given by Ohm's law, so that if a current $I$ is drawn from the battery we have the p.d. across the resistors $R_1$ and $R_2$ given as $V_1$ and $V_2$, respectively, where

$$V_1 = IR_1 \quad \text{and} \quad V_2 = IR_2$$

We now wish to replace the two resistors $R_1$ and $R_2$ by a single, effective resistor, $R$ say. By the term 'effective' we mean that we want the resistor to have the same effect on the circuit as the original resistors, i.e. we want this new resistor to draw the same current from the battery; see Figure 6.5. If $R$ is the combined or effective resistance and $V$ is the total p.d. across the resistors, then

$$V = V_1 + V_2$$

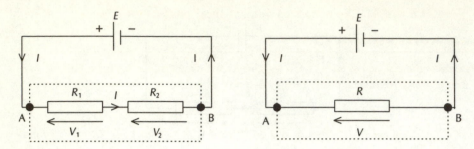

*Figure 6.5*    **Replace the resistors $R_1$ and $R_2$ by an effective resistor $R$**

This follows from the definition of the volt. Alternatively, an argument based on the fluid analogy in terms of pressure could be used.

Suppose we place a black box over the resistor $R$ and ask the question 'What is the resistance between terminals A and B?'. We could argue that the resistance is given by Ohm's law, (Equation (5.2)), i.e. by dividing the p.d. across the box, $V$, by the current flowing through the box, $I$. If we follow this suggestion, we find, upon substituting for $V$ in terms $V_1$, $V_2$ and, in turn, for $V_1$, $V_2$ in terms of $I$ and $R_1$, $R_2$ (namely $IR_1$ and $IR_2$, respectively), that the combined or effective resistance for two resistors $R_1$ and $R_2$ in series is given by the simple result

$$R = R_1 + R_2 \tag{6.2}$$

■■■ COMMENTS

- In this derivation it was very important that the currents flowing through the two resistors were the same, but that the p.d.s across each resistor were different.
- For two resistors in series the effective resistance is the sum of the individual resistances. It should be straightforward to derive from Equation (6.2) that for several resistors connected in series, the effective resistance is just the sum of the individual resistances.
- Since the same current flows through each resistor, the resistor with the largest value of resistance has the largest p.d. across its ends. It is straightforward to derive that, if $V$ is the p.d. across both resistors, then the p.d.s across $R_1$ and $R_2$ separately are given as $V_1$ and $V_2$, respectively, where

$$V_1 = \frac{R_1}{R_1 + R_2}\, V, \qquad V_2 = \frac{R_2}{R_1 + R_2}\, V$$

For this reason, resistors in series are sometimes known as 'potential dividers'.

- The total current flowing through the circuit depends upon both resistors, and is dominated by the largest resistance.

Worked Example 6.1 emphasizes some of these points:

**■■■ WORKED EXAMPLE 6.1**

Determine the p.d. across each of the resistors for the circuit shown in Figure 6.6(a).
− The effective resistance, $R$ say, is easily determined as

$$R = \text{sum of resistances} = 5\,\Omega + 10\,\Omega + 15\,\Omega = 30\,\Omega$$

so that the simplified circuit consists of a single resistor of resistance $30\,\Omega$ connected in series with the 5 V battery, as shown in Figure 6.6(b).

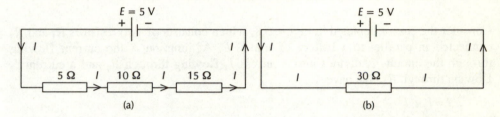

Figure 6.6    **(a) A circuit with three resistors in series. (b) The simplified circuit**

It is now straightforward to determine the current drawn from the battery. We find

$$I = \frac{V}{R} = \frac{5\,\text{V}}{30\,\Omega} = 0.167\,\text{A}$$

This current is the same as that drawn from the battery in Figure 6.6(a) (this is what we mean by the term *effective* resistance), so that we know the current flowing through each of the resistors in that figure. Given the current flowing through a resistor, the p.d. is determined from formula (5.1). We find

$$V_{(5\,\Omega)} = 0.833\,\text{V}, \qquad V_{(10\,\Omega)} = 1.667\,\text{V}, \qquad V_{(15\,\Omega)} = 2.500\,\text{V}$$

Note that the sum of the p.d.s adds up to the e.m.f. of the battery.

## 6.3  Parallel resistor network

Two resistors $R_1$ and $R_2$ are said to be connected in parallel if they are placed side by side with their corresponding ends joined together; see Figure 6.7. Note that in such an arrangement, the p.d.'s across the two resistors will be the same, but the current $I$ divides at the junction A such that a current $I_1$ flows through $R_1$ and a current $I_2$ flows through $R_2$. These currents merge at junction B to reform the original current $I$. The analogy of a fluid flowing through two parallel connected pipes suggests that this is reasonable.

We now proceed to derive the required result. The following derivation may be omitted, but the result (6.3) is important.

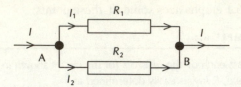

*Figure 6.7* **Two resistors connected in parallel**

○ ## Derivation of the result

Consider the circuit depicted in Figure 6.8, which consists of two resistors $R_1$ and $R_2$ connected in parallel to a battery of e.m.f. $E$. At junction A the current flowing through the circuit, $I$, divides into a current $I_1$ flowing through $R_1$ and a current $I_2$ flowing through $R_2$. We have

$$I = I_1 + I_2$$

Similarly, at junction B the currents $I_1$ and $I_2$ merge to form the current $I$ again:

$$I_1 + I_2 = I$$

The p.d.s across the two resistors are the same. The p.d. across a resistor is given by Ohm's law, (5.1), so that if a current $I$ is drawn from the battery we have

$$V = I_1 R_1 \quad \text{and} \quad V = I_2 R_2$$

We now wish to replace the two resistors $R_1$ and $R_2$ by a single, effective resistor, $R$ say. By the term 'effective' we again mean we want this resistor to have the same effect on the circuit as the original resistors, i.e. we want this new resistor to draw the same current from the battery.

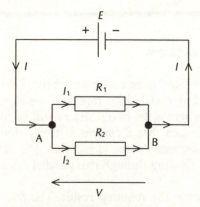

*Figure 6.8* **Two resistors connected in parallel to a battery**

Just as for the case of resistors in series, we can imagine placing a black box over the parallel resistor arrangement, and determining the effective resistance, $R$, of the box from (5.2), since we know the p.d. across the box, $V$, and the current flowing into the box, $I$. If $R$ is the effective resistance and $I$ is the total current flowing through the resistors then, as discussed previously,

$$I = I_1 + I_2$$

substituting for $I$ in terms of $V$ and $R$,

$$I = \frac{V}{R}$$

and $I_1$, $I_2$ in terms of $V$ and $R_1$, $R_2$, respectively,

$$I_1 = \frac{V}{R_1}, \qquad I_2 = \frac{V}{R_2}$$

we find that the combined or effective resistance for two resistors $R_1$ and $R_2$ in parallel is given as

$$\frac{1}{R} = \frac{1}{R_1} + \frac{1}{R_2} \tag{6.3}$$

## ▬ COMMENTS

- In this derivation it was very important that the p.d.s across the two resistors were the same, but the currents flowing through them were different.
- We sometimes write the result for $R$ as

  $$R = R_1 \parallel R_2$$

- For the case of two resistors $R_1$ and $R_2$ in parallel, the effective resistance can also be written in the form

  $$R = \frac{R_1 \times R_2}{R_1 + R_2} = \frac{\text{product of resistance}}{\text{sum of resistances}}$$

  This result is valid *only* for two resistors in parallel.
- The effective resistance $R$ is such that its value is less than that of either of the individual resistors

  $$R < R_1, R_2$$

- The result for several resistors in parallel can be obtained from the result (6.4) directly. We find that the inverse of the effective resistance is just the sum of the inverse of the individual resistances, i.e.

  $$\frac{1}{R} = \frac{1}{R_1} + \frac{1}{R_2} + \frac{1}{R_3} + \cdots$$

■ Since the p.d.s across the resistors are the same, more current flows through the smallest-valued resistor:

$$I_1 = I\,\frac{R_2}{R_1 + R_2}, \qquad I_2 = I\,\frac{R_1}{R_1 + R_2} \tag{6.4}$$

As a check, note that $I_1R_1 = I_2R_2$ ($= V$). This is reasonable, since the smallest-valued resistor, by definition, offers the least resistance to the flow of current.

## Special cases

■ If the two resistors are equal, i.e. if $R_2 = R_1$, then

$$R = \tfrac{1}{2}R_1$$

and the same current flows through each resistor.

■ If one of the resistors has a much smaller value than the other, say $R_1 \ll R_2$, then

$$R \approx R_1$$

i.e. the effective resistance is approximately given by the smaller-valued resistor. This is reasonable, since if $R_2$ is very much larger than $R_1$ most of the current will flow through $R_1$ and very little through $R_2$. Consequently, $R_2$ will only have a small effect on the circuit and can be ignored.

■ Suppose we place a piece of wire across a resistor. Taking the resistance of the wire to be zero, the resistance of this parallel arrangement is clearly zero. In such a case all the current flows through the wire and none through the resistor. We say that the wire has *shorted* the resistor.

Worked Example 6.2. shows one method of determining how the current splits at a junction.

━━━ **WORKED EXAMPLE 6.2**

Determine how the current splits at the junction for the circuit shown in Figure 6.9 (a).

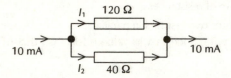

*Figure 6.9 (a)*

− Let us first determine the effective resistance, $R$ say. In a quite straightforward manner we find

$$R = \frac{\text{product of resistances}}{\text{sum of resistances}} = \frac{40 \times 120}{40 + 120} = 30\Omega$$

The 10 mA current flows through this effective resistor; see Figure 6.9(b).

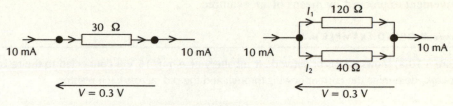

*Figure 6.9(b)* **Determine how the current splits at a junction**

It therefore follows that the p.d. across the effective resistor is

$$V = IR = (10 \times 10^{-3}\,\text{A}) \times (30\,\Omega) = 0.3\,\text{V}$$

But, according to the derivation, this is also the p.d. across the 40 $\Omega$ and the 120 $\Omega$ resistors in the original circuit (since they are in parallel). Given that we know the p.d. across each resistor, the current flowing through each one is easy to determine. We have

$$I_1 = \frac{V}{R_1} = \frac{0.3\,\text{V}}{40\,\Omega} = 7.5\,\text{mA}, \qquad I_2 = \frac{V}{R_2} = \frac{0.3\,\text{V}}{120\,\Omega} = 2.5\,\text{mA}$$

Why not just use the result (6.4)? There are two reasons for this: first, it is an extra formula to remember; and second, the method described above can be applied to any number of resistors in parallel, while the result (6.4) is limited to the case of two resistors in parallel. See Worked Example 6.3.

## 6.4 More general resistor networks

We now wish to consider resistor network circuits. As mentioned previously, all resistor networks of interest to us here can be made up of just two special arrangements of resistor networks:

*series* and *parallel*

We have previously discussed each of these networks separately. We now bring together these results.

**■■■ BASIC APPROACH FOR RESISTOR NETWORKS**

Using rules to be discussed, the resistor network of interest will be converted to a single resistor of resistance $R$, say. In this case the current drawn from the source of e.m.f. will be given by Equation (6.1). Using the same rules, we will be able to determine how the current $I$ divides and combines at the junctions of the original resistor network. Once the current flowing through a particular resistor has been determined, the p.d. across the resistor is given by Ohm's law, i.e. Equation (5.1).

We now continue with our discussion of resistor network problems. It is convenient to proceed by means of an example:

━━━ **WORKED EXAMPLE 6.3**

Figure 6.10(a) shows a resistor network. If a battery of e.m.f. 15 V is connected to the resistor network, determine the current flowing through and the p.d. across each resistor.

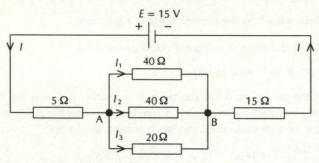

*Figure 6.10(a)*   **A resistor network problem**

− The first step is to determine the effective resistance of the resistor network. We first discuss the three resistors in parallel. We can either make use of the result (6.3) for two resistors in parallel and first determine the resistance of the two 40 Ω resistors in parallel, say $R_{(40 \,||\, 40)}$. We then have two resistors in parallel, the resistor $R_{(40 \,||\, 40)}$ and the 20 Ω resistor. Alternatively, we can make use of the generalization of (6.3) to three resistors in parallel. Either approach yields the effective resistance of the parallel arrangement as 10 Ω.

We have reduced the complexity of the resistor network shown in Figure 6.10(a) from five resistors to three resistors, 5 Ω, 10 Ω and 15 Ω, in series; see Figure 6.10(b). The effective resistance of the resistors in series is straightforward to determine, just being given as the sum of the individual resistances. Hence, the effective resistance of the network is given as 30 Ω.

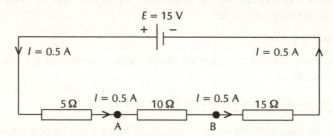

*Figure 6.10(b)*   **The network has been simplified**

We now have a single 30 Ω resistor in series with the 15 V battery. The current drawn from the battery is determined from Equation (6.1) as

$$I = \frac{E}{R} = \frac{15 \text{ V}}{30 \, \Omega} = 0.5 \text{ A}$$

Notice in Figure 6.10(b) that the same current, 0.5 A, flows through each of the 5 Ω, 10 Ω and 15 Ω resistors. Given the current flowing through a resistor, the p.d. is given by (5.1). In this example we have

$$V_{(5\,\Omega)} = 2.5 \text{ V}, \qquad V_{(10\,\Omega)} = 5 \text{ V}, \qquad V_{(15\,\Omega)} = 7.5 \text{ V}$$

We have determined both the currents flowing through and the p.d.'s across the 5 Ω and 15 Ω resistors. We have also determined the current flowing through and the p.d. across the 10 Ω resistor. This resistor was the effective resistance of the 40 Ω and the 20 Ω resistors in parallel. How are we to determine the current flowing through these resistors? This turns out to be straightforward. We make use of the fact that we know the p.d. across the parallel arrangement is the same as the p.d. across the effective 10 Ω resistor, i.e. 5 V (see Figure 6.10c). Thus, given

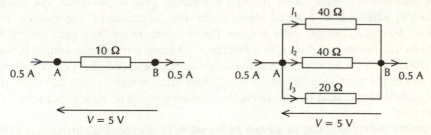

*Figure 6.10(c)*   **Determining how the current splits at the junction**

the p.d. across a resistor we can determine the current flowing through it from Equation (5.1). We find

$$I_1 = 0.125 \text{ A}, \qquad I_2 = 0.125 \text{ A}, \qquad I_3 = 0.250 \text{ A}$$

Note that the sum of these currents is 0.5 A, which is equal to the current flowing into (and out of) the junction, and that the largest current flows through the smallest resistor of the parallel arrangement.

---

As expected, the same current flows through each 40 Ω resistor and twice as much current flows through the 20 Ω resistor as through each of the 40 Ω resistors.

Note the basic approach: Simplify the circuit until we have one resistor in series with the battery, determine the current, then build the original circuit back up from the simplified circuit.

## 6.5   Power developed in an electric circuit

We first give a general definition of power: power is the rate at which work is done, i.e. the rate at which energy is used

$$\text{Power, } P = \frac{\text{energy used}}{\text{time taken}}$$

The SI unit of power is the watt (W), and is equal to the power used when one joule of work is performed in one second, i.e.

$$1 \text{ W} = 1 \text{ J s}^{-1}$$

For example, a 100 W lamp converts 100 J of electrical energy into heat and light each second, while a 750 W electric motor converts 750 J of electrical energy into mechanical energy (ignoring losses) each second.

## Dissipation of electrical energy

When an electric current flows through a resistor, heat is produced. As discussed previously, electrons in a metal move under the influence of the electric forces provided by the source of the e.m.f. As the electrons move they gain energy from these electric forces and their speed increases. Moving through the lattice, however, they collide with the atomic cores at the lattice sites. These collisions cause the atomic cores to vibrate more violently about their lattice sites, which is the equivalent of the metal getting hotter, heat being merely the random motion of atomic particles (recall Figure 5.7).

Consider the electric circuit shown in Figure 6.11. Suppose a current $I$ is flowing through the circuit and assume the p.d. across the resistor $R$ is measured as $V$. Then we quote the power developed across the resistor, given as

$$P = IV \tag{6.5}$$

We first note that (6.5) has the correct units since the units of $IV$ are

$$\frac{\text{C}}{\text{s}} \text{V} = \frac{\text{VC}}{\text{s}} = \frac{\text{J}}{\text{s}}$$

We now try to show that this is a reasonable definition. We know that current is the rate of flow of charge, i.e. if a charge $\Delta Q$ flows through the circuit in a time $\Delta t$ then, by definition (see Equation (3.1)),

$$I = \frac{\Delta Q}{\Delta t}$$

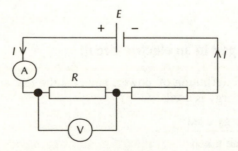

*Figure 6.11*   **The power developed across a resistor**

Substituting for the current $I$ in (6.5), we have

$$P = IV = \frac{\Delta Q}{\Delta t} V = \frac{\Delta Q \times V}{\Delta t}$$

But, from the definition of the volt (see Section 4.4), $\Delta Q \times V$ is just the amount of work done by the battery to move the charge $\Delta Q$ through the p.d., $V$, across the resistor $R$. Thus, we can interpret Equation (6.5) as

$$IV = \frac{\text{energy used to move charge across resistor}}{\text{time taken for charge to cross resistor}}$$

Thus the definition is reasonable in that it has the correct units and also gives the rate at which energy is used in moving the charge across the resistor.

There are several equivalent ways of writing down (6.5), all of which make use of Ohm's law (5.1):

$$\left.\begin{array}{l} P = IV = I(IR) = I^2R \\[2mm] P = IV = \dfrac{V}{R} V = \dfrac{V^2}{R} \end{array}\right\} \tag{6.6}$$

Resistors are usually rated with a 'power rating'. The power rating is the maximum power that can be dissipated by the resistor.

━━━ **WORKED EXAMPLE 6.4**

A 4.7 kΩ resistor has a power rating of 1/4 W. Determine the maximum current that can flow through and the maximum p.d. that can be developed across the resistor.

– Since we are given $P$ and $R$ it is convenient to make use of one of the results given in Equation (6.6), namely, $P = I^2R$. This will give the current directly. We are given $P = 1/4$ W and $R = 4.7$ kΩ $= 4700$ Ω; therefore

$$\tfrac{1}{4} = I^2_{max} 4\ 700$$

Thus,

$$I^2_{max} = \frac{1}{4} \frac{1}{4\ 700} \text{ A}^2$$

and so the maximum current that can flow through the resistor is $I_{max} = 7.29 \times 10^{-3}$ A $= 7.29$ mA. It follows that the maximum p.d. that can be developed across the resistor is 34.3 V.

━━━━━━━━━━━━━━━━━━━━━━━━━━━━━━━━━━

Note that the largest resistor does not necessarily dissipate the most power.

The heating effect is put to good use in, for example, the electric kettle. In an electric kettle the heating element is typically made from a material such as nichrome. This is an alloy of nickel and chromium which does not oxidize when the

━━ **WORKED EXAMPLE 6.5**

A 100 Ω resistor is rated at 3 W. Is it safe to connect it across a 20 V supply?
− The power dissipated by the resistor when connected across the 20 V supply is

$$P = \frac{V^2}{R} = \frac{20^2}{100} \text{ W} = 4\text{W}$$

This is larger than the permitted maximum and there is a possibility of the resistor burning out.

━━ **WORKED EXAMPLE 6.6**

Determine the total power dissipated by the circuit in Worked Example 6.3.

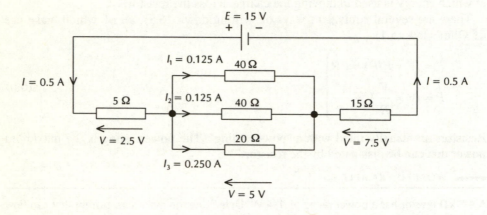

*Figure 6.12*

− Figure 6.12 shows the current flowing through and the p.d. across each resistor. It follows that we can determine the power dissipated by each resistor, being simply the product of the current through the resistor and the p.d. across it.

It follows that the power dissipated by the 5 Ω resistor is 1.25 W, each 40 Ω resistor is 0.625 W, the 20 Ω resistor is 5 W and the 15 Ω resistor is 3.75 W. This gives the total power dissipation by the circuit as 11.25 W.

material becomes hot. In other circumstances the heating effect represents useful energy which is lost; see, for example Section 10.4.

See the related discussion in Section 12.4.


## Internal resistance

Before proceeding further it is convenient to discuss the concept of 'internal resistance' for a source of e.m.f. Thus far, we have described a source of e.m.f.

■■■ **WORKED EXAMPLE 6.7**

What current flows through the heating element of an electric kettle rated at 2 kW given that the p.d. across the element is 240 V?
– The current flowing through the element is

$$I = \frac{P}{V} = \frac{2\,000}{240} \text{ A} = 8.333 \text{ A}$$

using one parameter, namely the size of the e.m.f., which indicates its ability to move charge around a circuit. A more realistic model includes the 'inbuilt' resistance to the flow of current, which is generally referred to as its 'internal resistance'. A more accurate representation of a battery is depicted in Figure 6.13, where the box is used to emphasize that the internal resistance, $r$, is inside the battery, and A and B are the terminals of the battery. The internal resistance of a source of e.m.f. acts like any other resistor in the circuit. The most important differences are that it is not possible, for example, to place a voltmeter across it as for a conventional resistor, or to remove the internal resistance from any circuit. Generally, we will ignore the box and take it as being understood.

Note that if you place a voltmeter across the ends of a source of e.m.f. this will cause a current to flow. Consequently, there will be a p.d. developed across the internal resistance and the voltmeter will not register $E$ but $E - Ir$. If we had a voltmeter that did not draw any current from the source, then $I = 0$ and the voltmeter would measure the e.m.f., $E$, of the source. For this reason, $E$ is sometimes referred to as the 'open-circuit voltage', the term 'open-circuit' implying that no current is drawn from the source.

## Maximum power transfer

Consider the circuit shown in Figure 6.14, which consists of a battery of e.m.f. $E$ and internal resistance $r$ connected to a resistance $R$. Let us determine the power developed across the resistor $R$. This is simply given by Equation (6.5), i.e. $P = IV$,

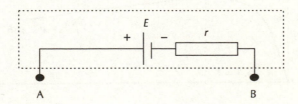

*Figure 6.13* **Internal resistance**

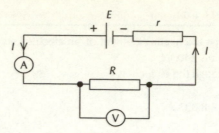

*Figure 6.14*   **The maximum power transfer theorem**

where $I$ is the current flowing through $R$, and $V$ is the p.d. across $R$. The current $I$ flowing through the circuit, and therefore through $R$, is easily determined as

$$I = \frac{E}{R+r}$$

since the resistors $r$ and $R$ are in series, the effective resistor which replaces them is $R+r$. It follows that the p.d. across $R$ is $IR$, i.e.

$$V = IR = E \frac{R}{R+r}$$

Given the current flowing through, $I$, and the p.d., $V$, across $R$, we can easily determine the power developed across $R$ as

$$P = E^2 \frac{R}{(R+r)^2}$$

We wish to choose $R$ to make the power developed across this resistor a maximum. Plotting the power, $P$, against the resistance, $R$, yields the graph shown in Figure 6.15. Note, that for small values of $R$, i.e. for $R \ll r$, the power developed across $R$ depends linearly on $R$, i.e. $P \propto R$, while for large $R$, i.e. for $R \gg r$, that the power developed varies as $1/R$, i.e. $P \propto 1/R$. Also, we note that the maximum power

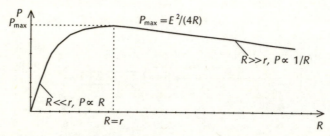

*Figure 6.15*   **The power developed across $R$ as a function of $R$ ($E$ and $r$ kept fixed)**

developed across $R$ occurs when $R$ is chosen equal to the internal resistance $r$, i.e. $R = r$. The maximum power developed across $R$ is

$$P_{max} = \frac{E^2}{4R}$$

## 6.6 Summary

The role of resistance in an electric circuit was discussed, the emphasis being on simple resistor networks connected to a single source of e.m.f. A general approach was discussed which could determine the currents flowing through and the p.d. across each resistor of the network.

This approach was to identify those resistors in the network which were connected in series and those connected in parallel. Simple rules were derived for replacing either two resistors in series or two resistors in parallel by a single effective resistor. We were then able to reduce any resistor network by a single resistor in series with the source of e.m.f., from which the current drawn could be determined. The way in which currents divide at junctions was used to determine the current flowing through any resistor in the network. Once the current flowing through a particular resistor was found, the p.d. across the resistor was given trivially from Ohm's law.

It was pointed out that when a current flows through a resistor, electrical energy is dissipated in the form of heat. The rate at which energy is dissipated was first quoted, and then discussed to show that the formula was reasonable.

## ▬ Problems

**6.1**   Two resistors connected in parallel have an effective resistance of 2.4 Ω. When connected in series, they have an effective value of 10 Ω. Determine the individual values.

**6.2**   Given the following values for five resistors:

$R_1 = 5\,\Omega,\quad R_2 = 10\,\Omega,\quad R_3 = 10\,\Omega,\quad R_4 = 100\,\Omega,\quad R_5 = 40\,\Omega$

determine the effective resistances of the resistor network shown in Figure P6.1, (a) between the points A and C, (b) between the points B and D, and (c) between the points A and D.

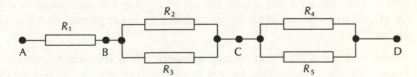

Figure P6.1

**6.3**   Consider the resistor network shown in Figure P6.2. If a battery of e.m.f. 10 V and negligible internal resistance is connected across points A and B, what currents flow in (a) the circuit? (b) the 10 Ω resistor? (c) the 30 Ω resistor? (d) the 120 Ω resistor? (e) Hence, determine the p.d. across each resistor.

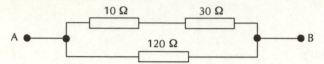

*Figure P6.2*

**6.4**   Determine the currents $I_1$, $I_2$ and $I_3$ in the circuit shown in Figure P6.3.

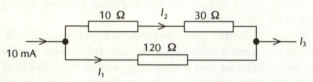

*Figure P6.3*

*Hint:* Replace the resistor network by a single equivalent resistor and determine the p.d. across this resistor. Hence determine the p.d. across each resistor. If the p.d. across each resistor is known, then the current flowing through it can be determined.

**6.5**   Consider the circuit shown in Figure P6.4, which shows a battery of e.m.f. 5 V and internal resistance 2 Ω connected to a 5 Ω resistor. Determine the current flowing through the circuit. An ammeter of resistance 1 Ω is placed in the circuit to measure the current flowing through the circuit. Indicate on a diagram how the ammeter is placed in the circuit and determine the current flowing registered.

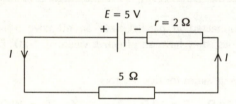

*Figure P6.4*

*Hint:* When the ammeter is connected into the circuit to measure the current, because of the non-zero resistance of the meter, the total resistance of the circuit changes. The current registered by the ammeter will be that flowing through the new circuit. Thus, the ammeter affects the circuit current it is attempting to measure. Ideally, the ammeter should be chosen such that it has a negligibly small effect on the circuit.

**6.6**   A battery of e.m.f. 20 V and internal resistance 2 Ω is connected as shown in the Figure P6.5. Determine (a) the current flowing through the circuit, and (b) the p.d. across each 40 Ω resistor. (c) A voltmeter of resistance 100 Ω is introduced in the circuit to determine the p.d.'s across the 40 Ω resistors. Indicate on a diagram how the meter is to be connected. What voltage will it indicate? Explain briefly why this is different from the value obtained in (b). (d) If the resistance of the voltmeter is 5 kΩ, what p.d. will it indicate?

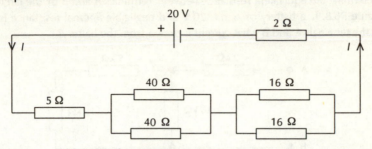

*Figure P6.5*

    *Hint:* When the voltmeter is connected into the circuit to measure the current, because of the non-zero resistance of the meter, the total resistance of the circuit changes. The p.d. registered by the voltmeter will be the product of the current flowing through the voltmeter and the meter resistance. Thus, the voltmeter affects the circuit current it is attempting to measure. Ideally, the voltmeter should be chosen such that it has a negligibly small effect on the circuit.

**6.7**    Making use only of the result for two resistors in parallel, i.e.

$$\frac{1}{R} = \frac{1}{R_1} + \frac{1}{R_2}$$

determine the resistance of the circuit shown in Figure P6.6 between terminals A and B.

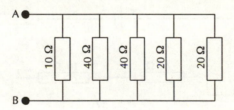

*Figure P6.6*

**6.8**    Consider the circuit shown in Figure P6.7, in which $V_{out} = 30$ V. Determine the value of the e.m.f., $E$, of the source.

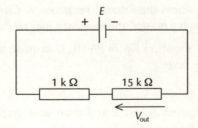

*Figure P6.7*

*Hint:* Given the p.d. across the 15 kΩ resistor it is possible to determine the current flowing through the circuit and, hence, through the 1 kΩ resistor.

6.9    Determine the equivalent resistance between terminals A and B of the circuit shown in Figure P6.8. If a battery of e.m.f. 20 V and negligible internal resistance is connected to the terminals A and B what current is drawn from the battery?

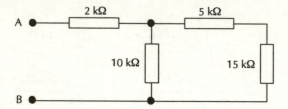

*Figure P6.8*

*Hint:* Redraw the circuit so that it is clearer which resistors are in series and which are in parallel.

6.10   A battery supplies a current of 0.6 A when connected to a 2 Ω resistor and a current of 0.2 A when connected to a 7 Ω resistor. Calculate the e.m.f. and the internal resistance of the battery.

6.11   Consider the circuit shown in Figure P6.9, which shows a battery of e.m.f. 5 V and internal resistance *r* connected to a resistance network. If the current drawn from the battery is 0.25 A determine the internal resistance of the battery.

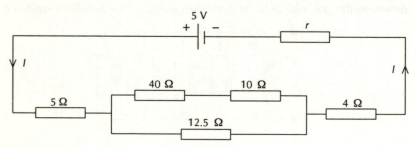

*Figure P6.9*

*Hint:* Determine the p.d.s across the 5 Ω, 12.5 Ω and 4 Ω resistors. From this determine the p.d. across the internal resistance *r*. Given the p.d. across and the current flowing through a resistor, the resistance may be determined.

6.12   Consider the circuit shown in Figure P6.10. Determine the power dissipated by the resistor *R* for *R* in the range

$$0 \leqslant R \leqslant 50 \, \Omega$$

Plot the power dissipated against *R* and from your graph determine the maximum power dissipated by *R*.

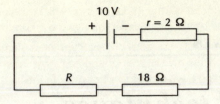

*Figure P6.10*

# CHAPTER SEVEN

# Circuit techniques

We previously considered simple resistor networks connected to a single source of e.m.f. and discussed a general approach to determine the currents flowing through and the p.d. across each resistor of the network. The basic approach was to identify those resistors in the network which were connected in series and those connected in parallel. Using simple rules for the effective resistance for two resistors in series or in parallel, together with a knowledge of how the current divides at junctions, it is possible to determine the current flow through and the p.d. across any particular resistor in the network. Essentially, circuits with one source of e.m.f. are solved in terms of Ohm's law.

We want to progress from simple networks, and, in particular, to consider circuits with two or more sources of e.m.f. In this situation, it is not obvious how Ohm's law can be used and we are forced to introduce several methods for these circuits. With these methods any circuit of arbitrary complexity can be solved in principle, though we will only consider relatively simple circuits which indicate the general approach of each method.

## 7.1 Kirchhoff's laws

We now discuss a method introduced by Kirchhoff which is able, in principle, to determine the currents flowing through any electrical circuit. We need to discuss two laws:

■■■ KIRCHHOFF'S CURRENT LAW

At any point in a circuit the algebraic sum of the currents directed into and out of the point must total zero.

■ **KIRCHHOFF'S VOLTAGE LAW**

> The algebraic sum of voltage sources and voltage drops (*IR* terms) must total zero around any closed path.

We discuss both laws in detail, discussing the physical reasoning behind them, and present examples illustrating their use. The examples will emphasize the importance of the term *algebraic*. We will find that in order to use Kirchhoff's laws successfully it is important to adopt conventions that determine the algebraic signs for the current and voltage terms.

### Kirchhoff's current law

The law is related to the more fundamental 'Law of Conservation of Electric Charge', which states that the amount of charges in a closed system is constant. In an electric circuit, if charge does not accumulate at any point in the circuit, then the amount of charge leaving a point must be the same as the amount of charge entering the point. Therefore, the *rate* at which charge leaves a point, which is just the current leaving the point, is the same as the rate at which charge enters that point.

We note that we have already made use of Kirchhoff's current law without explicitly stating it as such, when we derived the effective resistance for resistors in series and parallel in Sections 6.2 and 6.3, respectively. There, we justified the current relations we made use of by appealing to the analogy of a fluid flowing through a pipe.

#### Current convention

Take all currents directed into a point as positive and all currents directed away from that point as negative. Kirchhoff's current law is more readily appreciated in terms of examples.

■ **WORKED EXAMPLE 7.1**

Verify that Kirchhoff's current law is obeyed in Figure 7.1. The figure shows three resistors $R_1$, $R_2$ and $R_3$ carrying currents of 5 mA, 3 mA and 8 mA, respectively. This is a realistic use of Kirchhoff's current law which we will meet again later.

– For the point P we write

$$I_1 + I_2 - I_3 = 0$$
$$5 \text{ mA} + 3 \text{ mA} - 8 \text{ mA} = 0$$

Note that the signs on $I_1$, $I_2$ are taken as positive since these currents are directed towards point

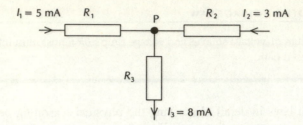

*Figure 7.1*   **Kirchhoff's current law. Notation: The point P is referred to as a node, branch point or junction**

P, while the sign on $I_3$ is taken as negative since it is directed away from point P. This is the significance of the term *algebraic* in the current law. A current with a positive value is taken to be directed towards a point, while a current with a negative value is taken as directed away from the point.

━━━ **WORKED EXAMPLE 7.2**

Determine the current $I_1$ in each of the two cases shown in Figure 7.2.

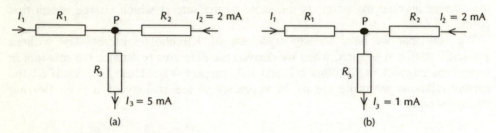

*Figure 7.2*   **Kirchhoff's current law**

− In Figure 7.2(a) the current directed to the point P is

$$I_{in} = I_1 + I_2 = I_1 + 2 \text{ mA}$$

while the current directed away from the point P is

$$I_{out} = I_3 = 5 \text{ mA}$$

Equating these quantities gives $I_1$ simply as

$$I_1 = 3 \text{ mA}$$

− In Figure 7.2(b) the currents directed to and away from point P are given as $I_1 + I_2$ and $I_3$, respectively. In this case we find

$$I_1 = -1 \text{ mA}$$

In this example $I_1$ has come out negative. The negative sign is readily interpreted as a current flowing in the opposite direction. Hence, the current $I_1$ is directed away from the point P. This indicates the algebraic nature of current in a circuit.

---

Taking account of the signs of the currents, we can write Kirchhoff's current law in the following manner:

$I_{in}$ (sum of currents into a point) $= I_{out}$ (sum of currents out of point)

In Worked Example 7.1 we have

sum of currents entering $P = I_{in} = I_1 + I_2 = 5$ mA $+ 3$ mA $= 8$ mA

sum of currents leaving $P = I_{out} = I_3 = 8$ mA

Worked Example 7.2 is instructive, emphasizing the importance of the term *algebraic*.

Kirchhoff's current law is the basis for the practical rule in parallel circuits that the total current flowing into the network must equal the sum of the branch currents.

Two equivalent ways of writing Kirchhoff's current law are:

> The algebraic sum of the currents entering and leaving any point in a circuit must equal zero

and

> The algebraic sum of currents into any point must equal the algebraic sum of the currents out of that point

### Kirchhoff's voltage law

From the definition of the volt (see Section 4.4 and Equation (4.1)), we see that Kirchhoff's voltage law tells us that around any closed loop, the sources of e.m.f. provide just enough work to move a charge across all the resistive elements.

#### Voltage convention

In determining the algebraic terms, we first have to mark the polarity of each source of e.m.f. and *IR* or potential drop term. The polarity of a voltage term indicates the direction in which any associated current will flow. In many practical examples we often make an assumption regarding the direction of current flow through the circuit. With this assumed choice of current flow, the direction of the voltage arrows across the resistors making the circuit is determined (recall the discussion in Section 4.3 and Figure 4.8). We now have to consider sources of e.m.f. and associate a consistent voltage arrow convention. Figure 7.3 shows the consistent voltage arrow notations adopted.

To justify this notation we appeal to a simple example consisting of a single source of e.m.f. in series with a resistor and show that the notation works; see

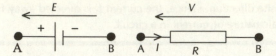

*Figure 7.3* **The voltage arrow notation adopted for sources of e.m.f. and p.d. terms**

Figure 7.4. The example also emphasizes the term algebraic as applied to Kirchhoff's voltage law.

In the method we must make a choice for the direction of the current. In the example under consideration the most obvious choice is take the current as indicated in Figure 7.4. Kirchhoff's voltage law states that the algebraic sum of voltage sources and voltage drops must total zero around any closed path. Let us consider the point A and walk around the circuit from this point, eventually returning to the point. We have two choices of direction in which to walk around the circuit: either a clockwise or a counter-clockwise direction. Suppose we arbitrarily decide to traverse the loop in a clockwise direction. The first voltage arrow we meet is that associated with the source of e.m.f. and it is of size $E$. The voltage arrow points in the *opposite direction* to which we are moving, so we associate with it a *minus* sign (this is where the term *algebraic* is important). Continuing our walk, we next meet the voltage arrow associated with the resistor. It is of size $V$ and the arrow points in the *same direction* we are travelling in, so we associate with it a *positive* sign. We continue our walk and eventually come back to the starting point, A. Applying Kirchhoff's voltage law to our walk around the closed loop, taking due care with the signs, we find

$$\text{algebraic sum of voltages} = -E + V = 0$$
$$\Rightarrow V = E$$

Substituting for the p.d. $V$ in terms of the current $I$ through the circuit, i.e. $V = IR$, yields the expected result:

$$I = \frac{E}{R}$$

i.e. the current through the resistor is equal to the e.m.f. of the source divided by the resistance.

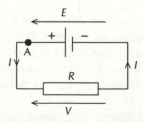

*Figure 7.4* **A single source of e.m.f. connected to a single resistor**

This is a rather long-winded derivation of an anticipated result. However, the example brings out several important points:

- The direction around the loop can be clockwise or counter-clockwise. In either case, if you come back to the starting point, the algebraic sum of all the voltage terms must be zero.
- If you do not come back to the starting point, then the algebraic sum found is the p.d. between the start and the finish points.

We made an assumption regarding the direction of the current flow. The choice made was reasonable and physically motivated. Worked Example 7.3 discusses this point.

▬▬▬ **WORKED EXAMPLE 7.3**

Apply Kirchhoff's voltage law to determine the current flowing in a circuit consisting of a single source of e.m.f. connected to a single resistor, making the assumption that the current flow in the circuit is from the negative terminal to the positive terminal; see Figure 7.5.

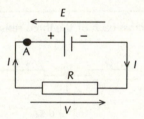

*Figure 7.5* **A single source of e.m.f. connected to a single resistor revisited**

– Notice in the figure the direction of the voltage arrows. The voltage arrow for the battery is, of course, the same as in Figure 7.4. However, the voltage arrow associated with the resistor has changed direction compared with that in Figure 7.4. This is related to the change in the assumed direction of the current. Starting from point A, arbitrarily traversing the loop in a clockwise direction, we have

$$\text{algebraic sum of voltages} = -E - V = 0$$
$$\Rightarrow V = -E$$

Substituting for the p.d. $V$ in terms of the current $I$ flowing through the circuit yields the result

$$I = -\frac{E}{R}$$

i.e. a negative current. A negative current is readily interpreted as a current flowing in the opposite direction to that given in the figure, so that the current comes out in the expected direction, i.e. flowing from the positive terminal to the negative terminal.

This example shows that if even if we make an incorrect assumption regarding the direction of the current flow in the circuit, by using Kirchhoff's laws consistently we still get the correct answer. This is an important point, since in examples with several sources of e.m.f. it is often not possible to know the direction of the current flow through the various circuit components beforehand, and we proceed by making assumptions regarding the directions of the current flow. The example suggests that this is not important – we merely have to interpret any negative currents we find as currents flowing in the opposite direction to that assumed.

The following is an equivalent way of writing Kirchhoff's voltage law:

> The algebraic sum of the voltages around any closed path is zero

Thus, if you start at any given point in the circuit at some given electric potential and move completely around the circuit in a closed path, coming back to the same point, the difference in potential must be zero.

We now present an example which makes use of both Kirchhoff laws. The example presents the basic procedure to be used in any Kirchhoff problem.

━━━ **WORKED EXAMPLE 7.4**

---

Determine the current and p.d. across each of the three resistors in the circuit shown in Figure 7.6(a).

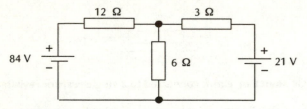

*Figure 7.6(a)*   **Two-source example**

– We begin by introducing currents $I_1$, $I_2$ and $I_3$ flowing in the circuit; see Figure 7.6(b). The directions of these currents may be chosen arbitrarily. The choices we have made for $I_1$ and $I_2$ are reasonable, since we know that if both sources of e.m.f. were not connected to each other, these are the directions in which we would expect the currents to flow. The choice for $I_3$ is made so that the application of Kirchhoff's current law is convenient. Once the directions of the currents are chosen, however, the direction arrows on the p.d.s are determined – since current flows from regions of high electric potential to regions of low electric potential.

Recall that Kirchhoff's voltage law talks about walking around closed paths, i.e. starting at some point in the circuit and walking around the circuit until we eventually end up back at the starting point. In this example, there are several such closed paths we can consider, for example ABEFA, ABCDEFA, ABEDCBEFA, etc. It turns out we only have to make use of the simplest and most obvious closed paths. In this example, the two obvious closed paths are ABEFA (call it loop 1) and CBEDC (call it loop 2). For convenience, the two loops are drawn separately in Figure 7.6(c). We stress that this is purely for convenience, and that the circuit we

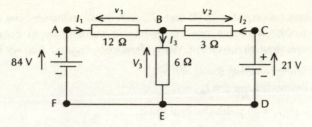

*Figure 7.6(b)* **Introducing the current flows and voltage arrows**

are solving is that shown in Figure 7.6(a). Figure 7.6(c) will simply make the application of Kirchhoff's laws easier.

Before proceeding further, let us make an important comment by considering, for example, loop 1 in Figure 7.6(c). In this figure we cannot replace the 12 Ω and 6 Ω resistors by a single 18 Ω resistor. These resistors are *not* in series, since different currents flow through each resistor (see Section 6.2). Remember, we have drawn the original circuit in terms of the separate loops purely for convenience.

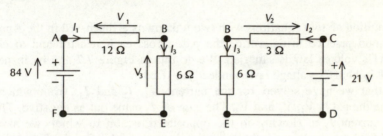

*Figure 7.6(c)* **Applying Kirchhoff's voltage law to each loop**

We now apply Kirchhoff's voltage law to each loop.

*Loop 1* Going around the loop in a clockwise direction

$$84 \text{ V} - V_1 - V_3 = 0$$
$$\Rightarrow V_1 + V_3 = 84 \text{ V}$$

From Ohm's law, we can relate the p.d. across a resistor to the current through it, i.e.

$$V_1 = 12 I_1 \quad \text{and} \quad V_3 = 6 I_3$$

Substituting for the p.d.s in terms of the currents $I_1$ and $I_3$ we have

$$12 I_1 + 6 I_3 = 84 \tag{1}$$

*Loop 2* Going around the loop in an counter-clockwise direction gives

$$21 \text{ V} - V_2 - V_3 = 0$$
$$\Rightarrow V + V_3 = 21 \text{ V}$$

Substituting for the p.d.s in terms of the currents $I_2$ and $I_3$ we have

$$3 I_2 + 6 I_3 = 21 \tag{2}$$

Thus far, we have two equations, (1) and (2), with three unknowns, the currents $I_1$, $I_2$ and $I_3$. To proceed further we make use of Kirchhoff's current law to obtain a relationship between the currents through the circuit. With reference to Figure 7.6(a) we have

sum of currents entering $B = I_{in} = I_1 + I_2$
sum of currents leaving $B = I_{out} = I_3$

Substituting for $I_3$ ($= I_1 + I_2$) yields the final result:

$$18\,I_1 + 6\,I_2 = 84$$
$$6\,I_1 + 9\,I_2 = 21$$

These two equations can be easily solved to give

$$I_1 = 5 \text{ A}, \quad I_2 = -1 \text{ A}$$

from which we can determine $I_3$ as

$$I_3 = I_1 + I_2 = 4 \text{ A}$$

The solution of two equations with two unknowns is discussed in the Appendix.

It is good practice to determine the p.d. across each resistor and to check that Kirchhoff's voltage law is satisfied. We do this in Figure 7.7 and by inspection we see that Kirchhoff's voltage law is indeed satisfied.

Note that we have solved for the currents $I_1$, $I_2$ and $I_3$, from which we can determine the p.d.'s $V_1$, $V_2$ and $V_3$. The sign of $I_2$ came out as negative. This means that the current $I_2$ is flowing in the opposite direction to which we assumed. In Figure 7.6(b) we made reasonable assumptions as to the flows of the currents, so how are we to explain the result for $I_2$? The result can be understood if we determine the p.d. across the 6 $\Omega$ resistor. We find

$$V_3 = 6 \times I_3 = 24 \text{ V}$$

This p.d. is larger than the e.m.f. of the battery, given as 21 V, and we therefore expect the current to flow as in the direction found, i.e. from a high electric potential to a low electric potential.

Recapping the basic steps we made, we have:

- Introduce the currents flowing in the circuit.
  The directions of these currents may be chosen arbitrarily, though it is best to make what appears to be reasonable choices if possible.
- Draw in the voltage arrows.
  Once the directions of the currents are chosen the direction arrows on the resistors are determined. The voltage arrows on the sources are fixed.
- Apply Kirchhoff's laws.

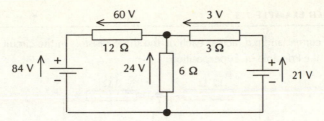

*Figure 7.7* **Checking the solution is correct is important. Note that we have reversed the direction of the voltage arrow on the 3 Ω resistor**

## 7.2 The principle of superposition

The Principle of Superposition is an alternative approach to the use of Kirchhoff's laws for the solution of circuits containing two or more sources of e.m.f. The method allows us to think of a complex circuit in terms of several simpler circuits, each of the new circuits consisting of only single sources of e.m.f. It is then straightforward to make use of Ohm's law to solve each of these simpler circuits. The Principle of Superposition relates all these simpler results to the solution of the original circuit. The theorem may be stated as follows:

▬▬ **THE PRINCIPLE OF SUPERPOSITION**

In any network consisting of resistors, the current flowing in any branch is the algebraic sum of the currents that would flow in that branch, if each source of e.m.f. were to be considered separately.

The theorem simply states that if we have several sources of e.m.f. in some circuit, the current flowing through a particular resistor can be thought of as being due to a current flowing due to the first source of e.m.f., with all others removed, added to the current flowing due to the second source of e.m.f., with all other sources of e.m.f. removed, until all the sources have been allowed for. The term *algebraic* simply means we must be careful to take into account that the separate sources may be causing currents to flow through the resistor in different directions.

Worked Example 7.5 reworks Worked Example 7.4, making use of the Principle of Superposition.

These results follow since, for example, the net current through the 12 Ω resistor consists of a current of 6 A due to the 84 V source and a current of 1 A, flowing in the opposite direction, due to the 21 V source. Thus, the current of −1 A flowing through the 3 Ω resistor is easily explained as the source $E_2$ pushing out a current of 3 A in the expected sense, and the source $E_1$ pushing a current of 4 A back through the source $E_2$.

━━━ **WORKED EXAMPLE 7.5**

Determine the current and p.d. across each of the three resistors in the circuit shown in Figure 7.8 in terms of the Principle of Superposition.

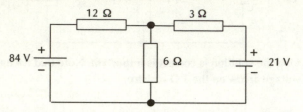

*Figure 7.8* **Two-source example in terms of the Principle of Superposition**

– According to the theorem we determine the current through a particular resistor in terms of the source $E_1$ separately and then determine the current due to the source $E_2$ alone. Figure 7.9(a) shows the circuit with source $E_1$ alone and source $E_2$ removed, while Figure 7.9(b) shows the circuit with source $E_2$ alone and source $E_1$ removed. The currents flowing in each circuit due to each source are also indicated. Since we are dealing with a circuit involving just one source of e.m.f. in each case, the choice of the directions of the current flow is straightforward.

It should be noted that in Figure 7.9(a), the 3 Ω and 6 Ω resistors are in parallel, while in Figure 7.9(b) the 6 Ω and 12 Ω resistors are in parallel.

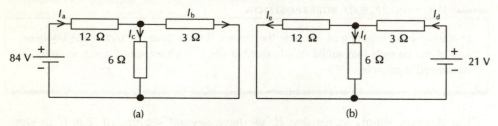

*Figure 7.9* **Applying the Principle of Superposition**

The Principle of Superposition has turned one difficult problem into two straightforward problems. Solving for the currents in both circuits should present no problem. We find

$$I_a = 6 \text{ A} \qquad I_d = 3 \text{ A}$$
$$I_b = 4 \text{ A} \qquad I_e = 1 \text{ A}$$
$$I_c = 2 \text{ A} \qquad I_f = 2 \text{ A}$$

Thus, the currents flowing through the three resistors (using the same notation as in Figure 7.6(b)) are given as

$I_1$ = net current through 12 Ω resistor = $I_a - I_e$ = (6 − 1) A = 5 A
$I_2$ = net current through 3 Ω resistor = $I_d - I_b$ = (3 − 4) A = −1 A
$I_3$ = net current through 6 Ω resistor = $I_c + I_f$ = (2 + 2) A = 4 A

which confirm the Kirchhoff calculation.

## 7.3 Thevenin's theorem and circuit models

Another method for solving complicated circuits that we wish to discuss is due to Thevenin. We begin by considering a lamp connected to a black box, as depicted in Figure 7.10.

If the lamp is lit what is the minimum we can say about the contents of the box? We must be able to argue that the box must contain at least a source of e.m.f. This follows since, if the lamp is lit, this implies that there must be a current through the lamp. If a current is flowing, there must be, at least, some source of e.m.f. inside the box. We actually go a little further than this minimum conclusion, and argue that, since all real sources of e.m.f. have an associated internal resistance, we can argue that the box is acting like a source of e.m.f. with an associated internal resistance. Thevenin made this observation somewhat more formal in the following theorem, which is depicted in Figure 7.11.

■■■■ **THEVENIN'S THEOREM**

> Any circuit consisting of many sources and components (no matter how they are interconnected) can be replaced by an equivalent series circuit with respect to any pairs of terminals in the network.

Thus, according to Thevenin, we can replace the effects of an arbitrarily complex network (merely some arrangement of several sources of e.m.f. and resistor combinations) by a single voltage source, the Thevenin voltage $V_{\text{Th}}$, in series with a single resistor, the Thevenin resistance $R_{\text{Th}}$.

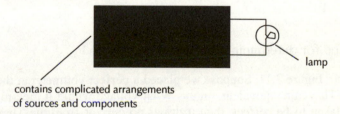

contains complicated arrangements
of sources and components

*Figure 7.10* **Beginnings of Thevenin's theorem**

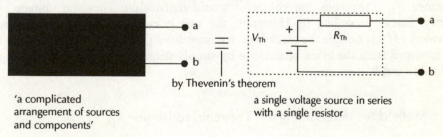

'a complicated
arrangement of sources
and components'

by Thevenin's theorem

a single voltage source in series
with a single resistor

*Figure 7.11* **Thevenin's theorem**

### The Thevenin voltage $V_{Th}$

The Thevenin voltage is the open-circuit voltage across the terminals a and b. The polarity of $V_{Th}$ is such that it will produce a current flowing from a to b, in the same direction as the original network.

### The Thevenin resistance $R_{Th}$

The Thevenin resistance $R_{Th}$ is the open-circuit resistance across the terminals a and b, but with all the sources shorted. This means that we determine the resistance by looking back into the network from the terminals a and b. The internal resistance is sometimes referred to as the open-circuit resistance.

Thevenin's theorem tells us that no matter how complicated the circuit is in the box in Figure 7.11 (for example, it may consist of several sources of e.m.f. connected to some complicated resistor network), we can represent the *effects* of this circuit on some particular component in terms of just one source of e.m.f. connected to one resistor. In terms of the example of the lamp connected to the black box in Figure 7.10, the Thevenin voltage $V_{Th}$ and the Thevenin resistance $R_{Th}$ are chosen such that the lamp shines just as brightly as with the original circuit.

This gives us a clue about how to choose the values for the Thevenin voltage $V_{Th}$ and the Thevenin resistance $R_{Th}$. These parameters, $V_{Th}$ and $R_{Th}$, must be chosen not only so this particular lamp has a particular brightness, but also if any other lamp is substituted for it, i.e. any lamp having a particular brightness in the original circuit must have the same brightness in the Thevenin equivalent circuit. Thus, $V_{Th}$ and $R_{Th}$ are determined by the insides of the black box, and not by the particular choice of the outside lamp.

### Method for determining the Thevenin voltage $V_{Th}$

Let us consider Figure 7.11. Suppose we placed a perfect voltmeter at the terminals a and b of the Thevenin equivalent circuit. What voltage would it measure? Since the voltmeter is taken to be perfect, then it draws no current from the Thevenin source of e.m.f. If no current flows, then there is no potential drop across the internal-like resistance $R_{Th}$. Therefore, the voltmeter would register the Thevenin voltage, $V_{Th}$ directly. Since, according to Thevenin, the two circuits in Figure 7.11 produce equivalent effects outside the black box, we must have that the Thevenin voltage $V_{Th}$ is determined from the black box circuit by simply determining the p.d. between the terminals a and b.

### Method for determining the Thevenin resistance $R_{Th}$

We consider Figure 7.11 again. To explain our approach for determining the

Thevenin resistance, $R_{Th}$, we first have to digress somewhat, and consider the operation of a device for measuring resistance – the ohm-meter:

### THE OHM-METER

We discussed on several occasions the use of an ammeter for measuring the current flowing through a resistor. Suppose we have a known source of e.m.f., $E$, of negligible internal resistance, connected to an ammeter as shown in Figure 7.12.

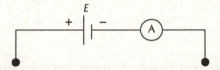

*Figure 7.12*    **The basic ohm-meter**

It is straightforward to note that if a resistor of resistance $R$, say, is connected across the terminals, then the ammeter will register a current of $E/R$. Since the source of e.m.f., $E$, is known, measuring the current means that the resistance $R$ can be determined. This is the basic principle of the ohm-meter: place a known p.d. across the resistor and determine the current.

One method to determine the Thevenin resistance, $R_{Th}$, is to place an ohm-meter directly across the terminals of the Thevenin equivalent circuit in Figure 7.11. Actually, we would be a bit more careful than this, since we have not yet allowed for the effects of the Thevenin voltage, $V_{Th}$. This would upset the operation of the ohm-meter. To determine $R_{Th}$ we would have to short-circuit $V_{Th}$, effectively removing the voltage source and replacing it by a piece of wire. If we followed this procedure on the Thevenin equivalent circuit, placing an ohm-meter on the terminals a and b would measure the Thevenin resistance, $R_{Th}$. Again, according to Thevenin, the two circuits in Figure 7.11 produce equivalent effects outside the black box, so we must have that the Thevenin resistance $R_{Th}$ is determined from the black box circuit by simply determining the resistance between the terminals a and b when we have short-circuited all voltage sources inside the black box, i.e. in effect replacing each voltage source by a piece of wire.

Note that if the voltage sources inside the black box have internal resistances associated with them, the internal resistances are to be treated as any other resistance. The term 'short-circuit the voltage sources' does not mean to remove any internal resistances.

Worked Example 7.6 illustrates these comments.

The approach we have described is an *operational* approach, i.e. it is the method we would use if we wanted to make use of voltmeters and ohm-meters. Another approach to the use of Thevenin's theorem is simply to quote the relevant steps.

━━━  **WORKED EXAMPLE 7.6**

Consider the circuit depicted in Figure 7.13(a). Make use of Thevenin's theorem to determine the p.d. across and the current flowing through the 12 Ω resistor.

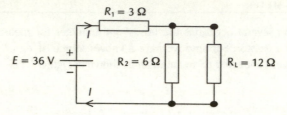

*Figure 7.13(a)*

– We are required to determine the p.d., $V_L$, across, and the current, $I_L$, flowing through the 12 Ω resistor (which in this context is often referred to as the load resistor, $R_L$). Although this problem can be trivially treated as a simple resistor network problem, it is very instructive to discuss it in terms of Thevenin's theorem.

The first step is to redraw the circuit so that we interpret the figure in terms of Figure 7.12, where $R_L$ replaces the lamp, and the rest of the circuit is the black box and is be thought of as a 'complicated arrangement of sources and components', see Figure 7.13(b).

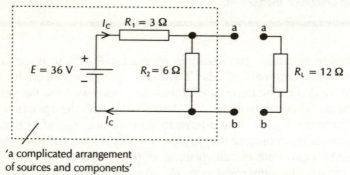

'a complicated arrangement
of sources and components'

*Figure 7.13(b)*  **Thinking of the problem in terms of Thevenin's theorem**

The next step is to apply Thevenin's theorem. This entails disconnecting the resistor $R_L$ and attempting to represent the circuit in terms of Thevenin's theorem; see Figure 7.13(c). Note in the figure that a current $I_C$ is drawn from the battery. This is not the same as the current drawn from the battery in the original circuit, the reason being that we have removed $R_L$, i.e. the circuit has changed. We interpret the p.d. across the 6 Ω resistor in terms of the Thevenin voltage, $V_{Th}$.

According to Thevenin, the effects of the two boxes at the terminals a and b are equivalent.

The current flowing through the circuit, $I$, is

$$I = \frac{36\ \text{V}}{(6+3)\ \Omega} = 4\ \text{A}$$

and the p.d. across the 6 Ω resistor is therefore

$$(4\ \text{A}) \times (6\ \Omega) = 24\ \text{V}$$

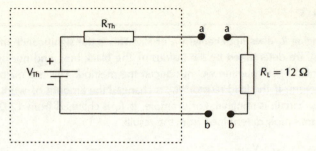

*Figure 7.13(c)* **According to Thevenin's theorem we can think of the circuit in terms of a single voltage source in series with a resistor**

giving the Thevenin voltage as

$$V_{ab} = V_{Th} = 24 \text{ V}$$

Notice in Figure 7.13(c) that, if we were to reconnect $R_L$ then current would flow through it from terminal a to terminal b. Comparing this with the Thevenin equivalent circuit, with the polarity of the Thevenin source chosen as in the figure, the current in this circuit would also flow from terminal a to terminal b.

We now have to determine the Thevenin resistance $R_{Th}$. First, short circuit the battery $E$ (the only source present) to find $R_{Th}$; see Figure 7.13 (d). Thus $R_{Th}$ is given as the effective resistance of $R$ and $R_2$ in parallel:

$$R_{Th} = 2 \text{ }\Omega$$

Note that we have taken the original figure and redrawn it as Figure 7.13(d), so that we can conveniently recognize that the resistors $R_1$ and $R_2$ are to be treated as parallel connected.

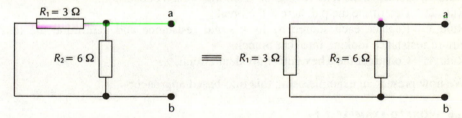

*Figure 7.13(d)* **Determining the Thevenin resistance**

Thus, the current flowing through and the p.d. across the load resistor $R_L$ are given as

$$I_L = \frac{V_{Th}}{R_{Th} + R_L}, \qquad V_L = I_L \times R_L$$

and we find

$$I_L = \frac{V_{Th}}{R_{Th} + R_L} = \frac{24 \text{ V}}{(2 + 12) \text{ }\Omega} = 1.714 \text{ A}$$

$$V_L = I_L \times R_L = 1.714 \times 12 \text{ V} = 20.571 \text{ V}$$

- Note that changing $R_L$ does not change $V_{Th}$ or $R_{Th}$. This is the significance of our comment that $V_{Th}$ and $R_{Th}$ are determined by the insides of the black box, and not by the particular choice of the outside lamp when we introduced the method. This is a major advantage of Thevenin's theorem: if the load resistor, $R_L$, is changed the amount of work we have to do to solve the new circuit is minimal. For example, if $R_L$ is changed from 12 Ω to, say, 20 Ω, then $V_L$ and $I_L$ are simply determined from the results

$$I_L = \frac{V_{Th}}{R_{Th} + R_L} = \frac{24 \text{ V}}{(2 + 20)\ \Omega} = 1.091 \text{ A}$$

$$V_L = I_L \times R_L = 1.091 \times 12 \text{ V} = 21.818 \text{ V}$$

i.e. the Thevenin part of the circuit remains unchanged if the load resistance is altered. If the circuit is solved as a resistor network problem, changing $R_L$ means solving the circuit from the beginning. Even in this simple example, the amount of work involved is not trivial.

- Note that the circuit we analyze apparently changes as we go through the different steps. For example, in the discussion above the resistors $R_1$ and $R_2$ are originally part of a resistor network problem with no particular relationship between them, and then they appear to be series connected resistors and finally parallel connected resistors. This is something that must be expected with the method and has to be accepted.

Careful reflection shows that the method we have described above can be summarized by the following four rules:

Rule 1:  Remove the load resistance from the branch concerned.
Rule 2:  Determine the p.d. across the break.
Rule 3:  Replace each source by its internal resistance and determine the total circuit resistance looking into that branch.
Rule 4:  Construct the Thevenin equivalent circuit.

We now present an example using this rule-based approach:

■■■ **WORKED EXAMPLE 7.7**

Determine the current flowing through the 5 Ω resistor in the circuit shown in Figure 17.14 (a).

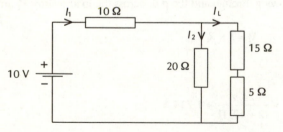

*Figure 7.14(a)*

*– Apply Rule 1:* Remove the load resistance from the branch concerned.
Application of rule 1 yields the circuit shown in Figure 7.14(b). Note that the current drawn from the battery has changed. This is because by removing the 5 Ω resistor we have necessarily changed the circuit under consideration.

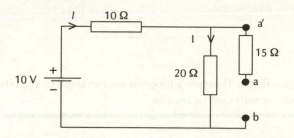

Figure 7.14(b)

*– Apply Rule 2:* Determine the p.d. across the break. Because we have removed the 5 Ω resistor this means that no current flows through the 15 Ω resistor. If no current flows through a resistor this means that there is no p.d. across the resistor. This implies that the points a and a' are at the same potential, so that the p.d. across the break is the same as the p.d. between the points a' and b. But the p.d. across the points a' and b is just the p.d. across the 20 Ω resistor. By inspection, the current flowing through the 20 Ω resistor is 1/3 A, so that the p.d. across it is 20/3 V. Thus we determine the Thevenin voltage as

$$V_{ab} = V_{Th} = \frac{20}{3} \text{ V} = 6.667 \text{ V}$$

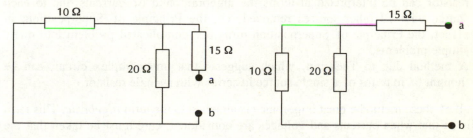

Figure 7.14(c)

*– Apply Rule 3:* Replace each source by its internal resistance and determine the total circuit resistance looking into that branch. We find we have to determine the resistance of the resistor network shown in Figure 7.14(c). The resistance of the network is easily determined as

$$R_{Th} = 15 + \frac{20 \times 10}{20 + 10} = 21.667 \ \Omega$$

since the 20 Ω and 10 Ω resistors in Figure 7.14(c) are in parallel.
*– Apply Rule 4:* Construct the Thevenin equivalent resistance. The Thevenin equivalent circuit is shown in Figure 7.14(d), from which the current is easily determined.

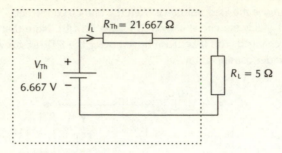

*Figure 7.14(d)*  **According to Thevenin's theorem we can think of the circuit in terms of a single voltage source in series with a resistor**

## 7.4 Summary

We previously considered simple resistor networks connected to a single source of e.m.f. and discussed a general approach to determine the currents flowing through and the p.d. across each of the resistors of the network. Essentially, circuits with one source of e.m.f. are solved in terms of Ohm's law.

Three approaches were described to treat more complicated circuits:

- a method that makes use of laws introduced by Kirchhoff;
- a method that makes use of the fact that a current flowing through a particular resistor can be interpreted in terms the algebraic sum of currents due to each source with all other sources removed, i.e. the Principle of Superposition; in effect, the Principle of Superposition turns one complicated problem into many simple problems;
- A method due to Thevenin, which suggests that any complex circuit can be thought of in terms of a single source in series with a single resistor.

In all of these methods, great importance is attached to the term *algebraic*. This term tells us that when currents and voltages are considered, care must be taken that the correct sign is associated with them.

## ▬▬ Problems

**7.1**    Consider the circuit shown in Figure P7.1. Use Kirchhoff's voltage law to determine the p.d.s. Given the following values for the resistors:

$R_1 = 10\ \Omega,\ R_2 = 160\ \Omega,\ R_3 = 140\ \Omega,\ R_4 = 120\ \Omega,\ R_5 = 10\ \Omega$

determine the current through each resistor. Confirm Kirchhoff's current law.

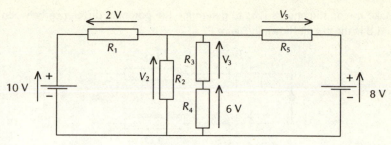

*Figure P7.1*

**7.2**    Consider the circuit shown in Figure P7.2. Draw in the respective voltage arrows for the *IR* drop terms and the source term. Determine the electric current, *I*, through the circuit. Comment on the sign of the current through the circuit.

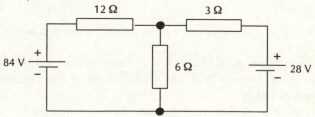

*Figure P7.2*

**7.3**    In the circuit shown in Figure P7.3 make use of Kirchhoff's laws to determine the current through each resistor. Confirm your results by solving the circuit making use of the Principle of Superposition.

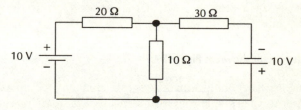

*Figure P7.3*

**7.4**    Consider the circuit shown in Figure P7.4. Make use of Kirchhoff's laws to determine the current through and the p.d. across each of the three resistors. Confirm your results by solving the circuit by making use of the Principle of Superposition. Comment on your results.

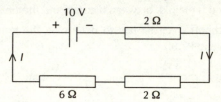

*Figure P7.4*

**7.5**    Make use of Kirchhoff's laws to determine the potential difference between points A and B in the circuit shown in Figure P7.5.

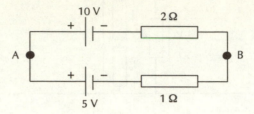

*Figure P7.5*

> *Hint:* Determine the current through the circuit. Then make use of Kirchhoff's voltage law and the result that if you do not come back to the starting point, then the algebraic sum found is the p.d. between the start and the finish points.

**7.6**    Make use of Kirchhoff's voltage law to determine the e.m.f. of the battery in the circuit shown in Figure P7.6.

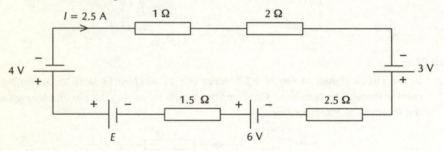

*Figure P7.6*

**7.7**    Consider the circuit shown in Figure P7.7.

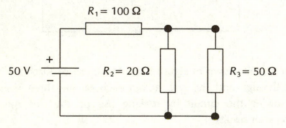

*Figure P7.7*

(a) Make use of Thevenin's theorem to determine the p.d. across and the current through $R_3$.

(b) Confirm your results by treating it as a resistor network problem.

(c) Make use of the results of part (a) to determine the current through $R_3$ if $R_3$ changes to 100 $\Omega$.

(d) What value of $R_3$ dissipates maximum power? Determine the maximum power dissipated across $R_3$.

**7.8** Consider the circuit shown in Figure P7.8. Determine the current through the resistor $R$ by means of Thevenin's theorem. Determine the Thevenin equivalent resistance, $R_{Th}$.

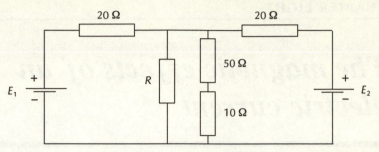

*Figure P7.8*

○ **7.9** Use Thevenin's theorem to determine the p.d. across and the current through the 6 Ω load resistor in the circuit shown in Figure P7.9. If the load resistor is changed to 12 Ω, determine the new current through $R_L$.

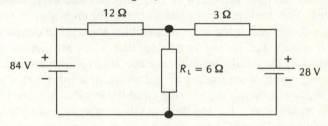

*Figure P7.9*

# CHAPTER EIGHT

# The magnetic effects of an electric current

Given that we have described electricity as electric charge in motion, much of the preceding work should have presented no real surprises and may, indeed, have even been anticipated. The analogy of a fluid flowing through a pipe makes concepts such as e.m.f., p.d. and resistance seem reasonable. Also, given our simple model of a conductor, it should not be too surprising that when an electric current flows through a conductor, the conductor will heat up. This follows from the observation that heat is merely the random motion of the atoms in a solid and when an electric current flows in, say, a metal, the moving conduction electrons can be imagined to collide with the atomic cores, forcing them to vibrate more about their lattice sites.

In the following we describe the magnetic effects of an electric current. This will be the beginnings of the description of a very important link between electricity and magnetism which was originally unsuspected. This connection will be developed over the next three chapters.

Historically scientists studied electricity and magnetism as separate subjects. It is now known there is a very important link between them. Indeed, this link is so substantial that, rather than talk about the subject of 'electricity and magnetism', the subject is given the special name of electromagnetism.

The course we will follow will, essentially, be that of merely describing experiments and interpreting them in the simplest possible manner. We begin by presenting a discussion of the magnetic effect of an electric current, first introducing the magnetic field via the familiar magnet.

## 8.1  Introduction to the magnetic field

As early as 600 BC the Greeks knew that a certain form of iron ore, now known as magnetite or lodestone, had the property of attracting small pieces of iron. Later, crude navigational compasses were made by attaching pieces of lodestone to wooden splints floating in bowls of water. These splints always came to rest pointing in a north–south direction, and were the forerunners of the familiar magnetic compass.

Suppose we take a magnet and place it under a piece of glass. If we lightly sprinkle iron filings over the glass it is found that they make a particular pattern, as can be seen in Figure 8.1. This is taken to indicate the pattern of the *magnetic field* associated with the magnet. We imagine that the magnet interacts with the iron filings via its magnetic field. We may define the magnetic field as the region around the magnet within which a force will be exerted upon magnetic materials.

The places on the magnet where the magnetic field appears to be most concentrated are referred to as the poles of the magnet. It is found that there are only two types of poles, which are conventionally termed N (north or north-seeking) and S (south or south-seeking).

If the N pole of one magnet is brought near the N pole of another magnet, it is found experimentally that the two poles repel each other. Similarly, if the two S poles are brought close to each other, then the two poles repel each other. However, when a N pole is brought close to a S pole it is found that the two poles attract each other. These results may be conveniently summed up in the law

like poles repel, unlike poles attract.

This sounds very similar to our discussion of electric charges, where we found, for example, that

only two types of electric charge

like charges repel, unlike charges attract.

Thus, it is tempting to argue that N and S are two types of 'magnetic charge', which are different in some manner. An important difference is that, whereas we can talk about isolated positive and isolated negative charges, isolated magnetic poles have never been observed., i.e. we only ever deal with N–S arrangements.

Thus, the reason why a compass aligns itself in a north–south direction is accounted for if we argue that the Earth itself acts like a magnet. With the rules that like poles repel and unlike poles attract each other, we note that the magnetic field of the Earth must act such that there is a magnetic south pole at the geographic north pole and, similarly, there is a magnetic north pole at the geographic south pole. Thus, a compass aligns itself such that its N pole points towards the magnetic south pole and, similarly, its S pole points towards the magnetic north pole of the Earth.

## The magnetic field

We now wish to discuss the pattern that the iron filings take in Figure 8.1. As stated previously, this pattern is taken to indicate the pattern of the magnetic field associated with the compass. We generally represent any magnetic field in terms of lines of magnetic force or lines of magnetic flux. Note that the iron filings take the same pattern as the magnetic field because they respond to, or are subject to the

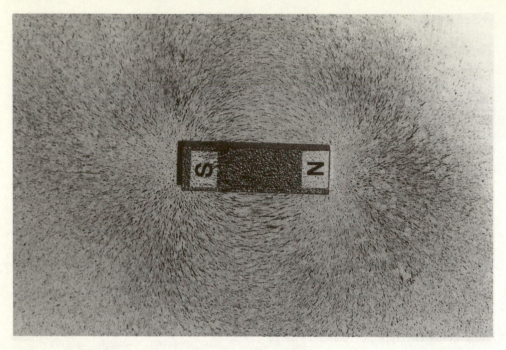

*Figure 8.1* **The pattern of iron filings around a magnet**

effects of, magnetic forces, and the magnetic field exists independently of whether the iron filings are present or not.

A magnetic field is an example of a vector quantity, and as such two quantities are required to fully describe it:

> its direction and its strength.

### The direction of the magnetic field

The direction of the magnetic field is taken to be the same as the direction in which ɹ free magnetic N pole would tend to move in the field, i.e. the direction of the *force* which acts upon a free N pole. Since we only find N–S poles in pairs, the direction of the magnetic field is taken to be the same as the direction which the N pole points to (which is opposite to the direction the S pole points to); see Figure 8.2.

### The strength of the magnetic field

We can investigate the strength of a magnetic field by placing a small compass in the field; the small compass is said to act like a probe. Because the compass experiences magnetic forces, it will tend to align itself with the magnetic field as indicated previously. The strength of the magnetic field tells us how difficult it is to

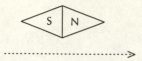

*Figure 8.2*   **A line of magnetic field with the direction of the field indicated by an arrow. The direction of the field is taken to be the same as the direction in which the N pole of a compass points when placed in that field**

move the probe compass from its aligned position. We can, loosely, think of the strength of the magnetic field in terms of the *density* of the lines of magnetic flux.

Thus, where the lines of magnetic flux are close together we expect the field to be stronger than at other points where the lines of magnetic flux are further apart.

### The SI unit of the strength of the magnetic field

The strength of the magnetic field is usually given the symbol $B$. The term $B$ is taken to indicate the strength of the magnetic field at a particular point, and is often referred to as the *magnetic flux density*. The SI unit of magnetic field strength or flux density, $B$, is the tesla (T).

The strength of the magnetic field can be thought of as measuring the number of lines of magnetic force, or lines of magnetic flux, per unit area. An equivalent SI unit is the weber per square metre (Wb m$^{-2}$)

$$1\ T \equiv 1\ Wb\,m^{-2}$$

### Properties of lines of magnetic force or flux

We have argued that we can represent a magnetic field by lines of magnetic force or flux. The arrow on a line of flux indicates the direction in which a compass would align itself. To explain the experimental observations of how magnets behave it is necessary to attribute these lines of flux with certain properties:

- Lines of flux always form closed loops, beginning on a N pole and finishing on a S pole.
- Lines of flux cannot cross one another.
- Lines of flux take the shortest possible path.
- Parallel lines of flux in the same direction repel one another.

For completeness, we point out that lines of magnetic flux always form closed loops; this is related to the absence of isolated N or S magnetic poles.

We use these properties to suggest that like poles repel and unlike poles attract; see Figure 8.3(a) and (b), respectively. Figure 8.3(a) shows the magnetic field for two magnets with their N and S poles adjacent to each other. The lines of flux go from the N pole of one magnet to the S pole of the other magnet. Because the magnetic lines of flux repel each other, very few flux lines follow the shortest, most direct path.

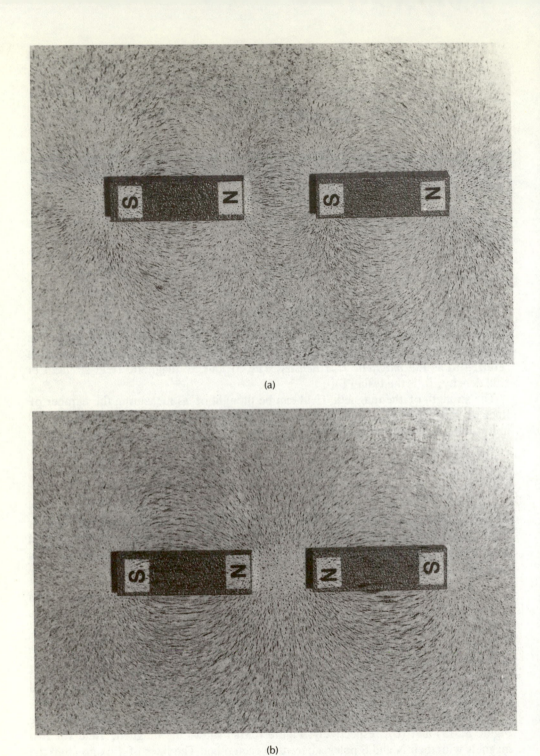

(a)

(b)

*Figure 8.3* **The attraction and repulsion of unlike and like poles explained in terms of magnetic lines of flux**

Since flux lines always attempt to minimize their length, there is an attractive force between the two magnets. When two like poles are adjacent to one another, as in Figure 8.3(b), no flux lines go from one magnet to the other. The flux lines from each pole repel one another and a repulsive force is exerted between the two magnets.

## 8.2   The magnetic field due to the flow of an electric current

We now begin our discussion of the connection between electricity and magnetism. Consider the following experiment, depicted in Figure 8.4. A compass is placed near a conductor, the compass pointing in some arbitrary direction. We assume that there are no other sources of magnetic field close by, so the compass experiences no magnetic forces. This means that if we move the compass from the direction it happens to be pointing in, to another direction, then the compass will remain in the new direction. It is only in the presence of a magnetic field that the compass experiences magnetic forces. Note that since the switch is open, there is no current flowing through the circuit.

Now consider what happens when we close the switch (see Figure 8.5). We find that the compass is deflected from its original position, as indicated in Figure 8.5. How are we to explain this? The simplest explanation of this experimental observation is to argue that a conductor carrying an electric current creates a magnetic field. This follows since the compass, which is merely a small magnet, responds to the presence of magnetic fields. The magnetic field created by the electric current interacts with the magnetic field of the compass to produce a force acting on the compass – i.e. the compass will experience a (magnetic) force and will therefore tend to move or be deflected.

Note that we have not explained *how* or *why* an electric current creates a magnetic field. All we know is that the compass is deflected. Since the compass responds to magnetic fields, we must have that the current creates a magnetic field. There is little else we can suggest as an explanation.

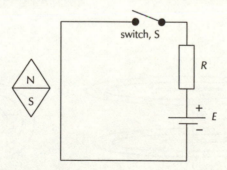

*Figure 8.4*   **The compass has been placed in an arbitrary position. The switch is open and no current flows through the circuit**

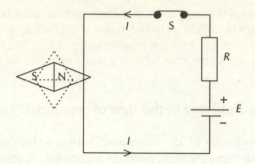

*Figure 8.5* **The switch is closed and the compass is deflected**

Given there is a magnetic field associated with an electric current there are two properties of this field that interest us

the pattern of the field and what the strength of the field depends upon.

The pattern of the magnetic field depends on the path the current takes. We discuss three cases of current flow.

## The magnetic field due to a long straight wire

It is found that the simplest pattern is for a long straight wire. The lines of magnetic flux in this case are found to take the form of concentric circles. This can be verified by placing a small compass near a wire carrying a current and observing the deflection as the compass is moved around the wire; see Figure 8.6.

A convenient way of remembering the direction of the magnetic field due to an electric current is given by the

'right-hand grip' rule

According to this rule, the thumb of the right hand points along the direction of the current and the direction of the fingers give the direction of the magnetic field; see Figure 8.7.

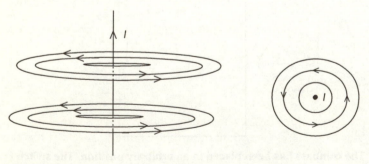

*Figure 8.6* **The pattern of the magnetic field for a long straight wire**

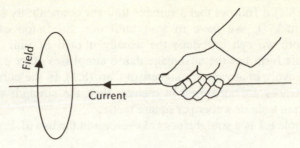

*Figure 8.7* **The right-hand grip rule (thumb – current; fingers – field)**

Having discussed the direction or the pattern of the magnetic field created by the current, we now have to discuss its strength, $B$. Experimentally we find that the strength of the magnetic field for a long thin wire depends upon two factors:

the amount of current flowing and the distance from the wire

We might expect that as we increase the current the strength of the magnetic field at any given point will increase. Experiments show that the magnetic field strength at a given position depends linearly on the current through the wire

$B \propto I$

We may also expect that as we move further away from the wire, the strength of the field gets weaker. This is borne out by experiments, which show that for a given current flowing through the conductor, the magnetic field strength varies inversely to the distance from the conductor

$$B \propto \frac{1}{r}$$

Combining these two results, we have

$$B \propto \frac{I}{r}$$

As usual we turn a proportionality into an equality by introducing a constant. It is conventional to write

$$B = \mu_0 \frac{I}{2\pi r} \tag{8.1}$$

The constant $\mu_0$ is referred to as the permeability of free space and measures the ease with which magnetic forces are transmitted through space. It has the numerical value in SI units of

$$\mu_0 = 4\pi \times 10^{-7}$$

From Equation (8.1) it follows that a suitable unit for permeability is the tesla metre per ampere $(T\,m\,A^{-1})$. We have to wait until our discussion of inductance in Chapter 10 before we can introduce the usually quoted SI unit. Introducing the constant $\mu_0$ in this form has the advantage that it simplifies other equations that are used more extensively in advanced magnetism than (8.1). If the current is measured in amperes, distances are measured in metres, then the strength of the magnetic field, $B$, is given in tesla or weber per square metre.

Worked Example 8.1 is a straightforward exercise in the use of Equation (8.1):

### ■■■■ WORKED EXAMPLE 8.1

A current of 100 mA flows through a long straight conductor. Determine the strength of the magnetic field and the direction in which the N pole of a small compass would point if placed at the points X, Y and Z in Figure 8.8.

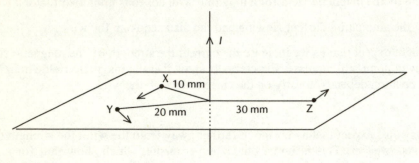

*Figure 8.8*

– For example, at the point X, a distance of 10 mm from the wire, the strength of the magnetic field is

$$B(X) = 2 \times 10^{-7} \times \frac{100 \times 10^{-3}}{10 \times 10^{-3}}\ T = 2 \times 10^{-6}\ T$$

where we have conveniently written $\mu_0/(2\pi)$ as $2 \times 10^{-7}$. Similarly, the strength of the field at the points Y and Z are given as

$$B(Y) = 1 \times 10^{-6}\ T \qquad B(Z) = 0.667 \times 10^{-6}\ T$$

The directions of the field at these points are given by the right-hand grip rule and are indicated in Figure 8.8 as the arrowhead line segments, the lengths of the lines being proportional to the strength of the field.

## The magnetic field due to a short coil of wire

We have previously seen that the strength of the magnetic field around a long straight wire is related to the current flowing through the wire, $I$, and the distance, $r$, from the wire by the result

$$B = \mu_0 \frac{I}{2\pi r}$$

while the pattern of the magnetic field is shown in Figure 8.6.

We first consider the magnetic field due to a single circular loop. What does the pattern of the magnetic field look like? Figure 8.9 shows the result for the magnetic field viewed from above. Note that very close to the wire, the pattern of the field is approximately that of concentric circles, as we would expect for isolated straight wires. This gives a clue to what is happening. We can think of the magnetic field as being due to the result of the magnetic fields of two wires carrying currents in opposite directions. At points near the centre, the magnetic fields of these two wires add or superpose to make the total magnetic field stronger than that due to one piece of wire alone.

When the same wire carrying the same current is bent into a circle of radius $r$, the strength of the field at the *centre* of the loop is given as

$$B = \mu_0 \frac{I}{2r}$$

Since $\pi \approx 3$, simply bending the wire into a loop increases the strength of the magnetic field (at the centre of the loop) by a factor of approximately 3.

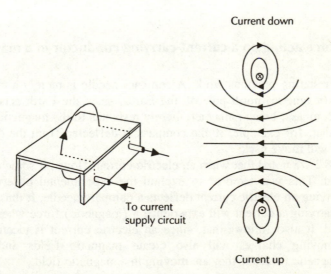

**Figure 8.9    The pattern of the magnetic field for a single circular loop**

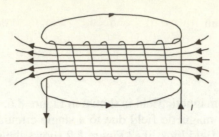

*Figure 8.10*   **The pattern of the magnetic field for a long coil of wire**

## The magnetic field due to a long coil of wire

Let us now consider the magnetic field for a long coil, i.e. for a coil which is long compared with its radius. Figure 8.10 shows the magnetic field due to a current flowing through the coil. As indicated in the figure, the magnetic field inside the coil is approximately uniform and is

$$B = \frac{\mu_0 n I}{l} \tag{8.2}$$

where $l$ is the length of the coil and $n$ is the number of turns.

The term uniform means not only that the magnetic field has the same strength at each point inside the coil, but also that the field points in the same direction at each point inside the coil.

## 8.3   The force acting on a current-carrying conductor in a magnetic field

We begin by restating previous work: A compass needle is merely a magnet which points to the (magnetic) north pole of the Earth, since the Earth acts like a giant magnet. The compass, being a magnet, merely responds to the magnetic field due to the Earth so that, for example, if the compass is deflected from the direction it is pointing in, it will move back.

In Section 8.2 we noted that when an electric current flows, it somehow creates a magnetic field. This was required to explain the experimental observation that a conductor carrying an electric current deflects a compass needle. It thus follows that a conductor carrying a current will experience a (magnetic) force when placed in a magnetic field. It also follows that, since an electric current is electric charge in movement, moving charges will also create magnetic fields and, therefore, experience magnetic forces if they are moving in a magnetic field.

Consider a current in a piece of conductor, with the conductor placed in the magnetic field of some permanent magnet: see Figure 8.11. The magnetic field is supplied by a shaped permanent magnet. The magnet is shaped in such a manner that the magnetic field between the poles is uniform. This is for experimental convenience.

The force acting on the conductor has two parts:

>     direction and magnitude

The direction tells us which way the conductor tends to move, and the magnitude tells us the strength of the force exerted on the conductor.

Strictly, the force only acts on the moving charges which make up the electric current, but, because they (i.e. the conduction electrons for the case of a metal conductor) are bound to the conductor, the whole conductor will move. The origin of this force can be explained in terms of the properties attributed to the lines of magnetic flux discussed earlier. Figure 8.12(a) shows the uniform magnetic field of the permanent magnet, while Figure 8.12(b) shows the magnetic field due to the electric current alone. Using the properties of magnetic lines of flux, the resultant field is shown in Figure 8.12(c).

Note that in region X in Figure 8.12(c), the magnetic lines of flux combine to give a resultant field in which the lines of flux are in the same direction, while in the region Y, the resultant field is weaker, since the fields due to the permanent magnet and the current are in opposite directions. Making use of the property that magnetic flux lines try to repel each other, we expect the force on the wire to be in the direction indicated.

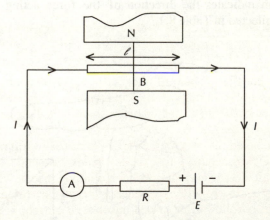

*Figure 8.11* **The force acting on a conductor carrying a current *I* placed in a magnetic field *B***

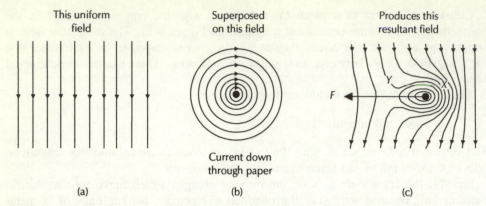

This uniform field

Superposed on this field

Produces this resultant field

Current down through paper

(a)                    (b)                    (c)

*Figure 8.12* **Interpreting the force in terms of the properties attributed to the lines of flux**

A convenient way of remembering the direction in which the force acts is given by Fleming's left-hand rule, while the magnitude of the force is determined from experiments. We now discuss each of these in turn.

### Direction of the force

The direction of the force may be determined by making use of Fleming's left-hand rule; see Figure 8.13. When using this rule, the first finger, second finger and thumb of the left hand are placed in mutually perpendicular positions as indicated in Figure 8.13. The first finger is taken to point in the direction of the magnetic field, while the second finger points in the direction of the current flowing in the conductor. The direction the thumb indicates the direction of the force acting on the conductor. These results are collected in Table 8.1.

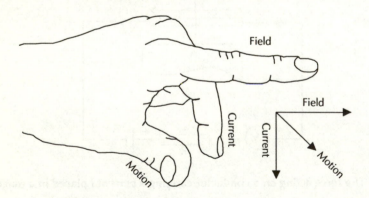

Field

Field

Current

Current

Motion

Motion

*Figure 8.13* **Fleming's left-hand rule**

*Table 8.1* **Fleming's left-hand rule**

| | |
|---|---|
| First finger | points in the direction of the magnetic **F**ield |
| Se**C**ond finger | points in the direction of the **C**urrent flow |
| Thu**M**b | points in the direction of the **M**otion of the conductor |

## Magnitude of the force

We expect the force to depend on the following factors:

- The strength of the magnetic field of the permanent magnet. This follows since we have indicated that the force on the conductor is due to the magnetic field associated with the current flowing through it and its interaction with the magnetic field of the permanent magnet. We expect that if we increase the field of the permanent magnet, then the (magnetic) force will increase.
- The amount of current flowing through the conductor. Recall, the magnetic field associated with an electric current was found to be proportional to the size of the current.
- The length of the conductor in the magnetic field. We might expect that as the current flows through the conductor, it experiences a force due to the cutting of more lines of magnetic flux of the permanent magnet. If the length of the conductor through which the current flows increases, the more lines of magnetic flux will be cut and the total force would be expected to increase.

Experimentally, we find that the magnitude of the force, $F$, is proportional to the product of these three factors:

$$F \propto BIl$$

If we use SI units for the quantities on the right-hand side of the above equation, i.e. the (permanent) magnetic field is expressed in tesla, the current flowing through the conductor in amperes and the length of the conductor in the magnetic field in metres, then the force, $F$, as measured in newtons is given as

$$F = BIl \qquad (8.3)$$

━━ **COMMENTS**

- The result for the force given above is for the experimental arrangement indicated in Figure 8.11, where the current 'cuts' the lines of magnetic field.
- This is the maximum force possible. If the conductor lies in a direction which is parallel to the magnetic field of the permanent magnet, the force is zero. In general, the current $I$ which appears in (8.3) is the component which flows perpendicular to the magnetic field; see Figure 8.14.

Thus, the force acting on the conductor of length $\ell$, carrying a current $I$ flowing at an angle to the magnetic field, $B$, is given as

$$F = BI_{\text{eff}}\ell = BI\ell \sin(\theta) \qquad (8.4)$$

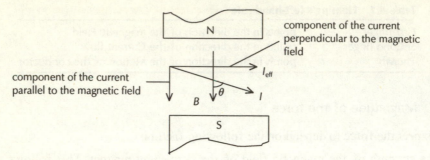

component of the current perpendicular to the magnetic field

component of the current parallel to the magnetic field

*Figure 8.14*   **The effective current $I_{eff}$**

since the component of the current perpendicular to the magnetic field is

$$I_{eff} = I \sin(\theta)$$

As special cases we recover (8.3) if we put $\theta = 90°$, i.e. the current is perpendicular to the magnetic field or cuts the lines of magnetic field, while if we put $\theta = 0°$, i.e. the current flows in a direction along the lines of magnetic field, then the force is zero.

▬▬ **WORKED EXAMPLE 8.2**

Determine the force, i.e. magnitude and direction, on the conductor placed in a magnetic field, as shown in Figure 8.15

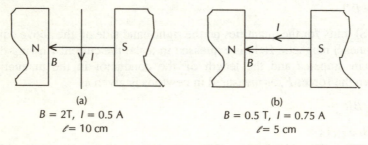

(a)
$B = 2T, I = 0.5 A$
$\ell = 10$ cm

(b)
$B = 0.5$ T, $I = 0.75$ A
$\ell = 5$ cm

*Figure 8.15*

In Figure 8.15(a) the force is given as

$$F = BII = 2 \times 0.5 \times 0.1 \text{ N} = 0.1 \text{ N}$$

since the current cuts the lines of magnetic field, i.e. the magnetic field and the current are perpendicular. The direction of the force is given by Fleming's left-hand rule. The force acts in a direction into the plane of the figure. In Figure 8.15(b) the force is zero since the current flows in a direction along the lines of magnetic field.

## 8.4 The magnetic force between two current-carrying conductors

An important special case to consider is the force between two current-carrying parallel conductors. Suppose we have two thin, very long conductors which we place a distance $r$ apart. Suppose also that we arrange for a current $I_1$ to flow through the first conductor and a current $I_2$ to flow through the second conductor; see Figure 8.16.

Since currents are flowing in both conductors there will be magnetic fields associated with each current. We therefore expect some magnetic interaction between the two conductors. We begin by considering the magnetic field created by the first conductor carrying a current $I_1$ at a distance $r$ from it. As explained previously, the magnetic field forms concentric circles about the conductor (recall Figure 8.6.).

It follows that if the second conductor is thin, at this point it will experience a uniform magnetic field, $B_1$, along the entire length of the conductor, i.e. over the thickness of the second conductor, the magnetic field ($B_1$) does not vary significantly; see Figure 8.17.

Now consider a length $l$ of the second conductor. As far as the second conductor is aware, it is in a uniform magnetic field of strength $B_1$, in a direction as indicated in the figure (recall that we assumed that both conductors are thin).

Since the current $I_2$ and the magnetic field $B_1$ are mutually perpendicular, the magnitude of the force acting on a length $l$ of the second conductor is determined from Equation (8.3), and is given as

$$F_{2,1} = B_1 I_2 l$$

where $F_{2,1}$ is the force acting on the second conductor carrying a current $I_2$, due to the magnetic field associated with the current $I_1$ flowing in the first conductor.

If we substitute for $B_1$ in terms of $I_1$ and the separation $r$ we find we can express $F_{2,1}$ in the following manner:

$$F_{2,1} = \mu_0 \frac{I_1 I_2 l}{2 \pi r} \tag{8.5}$$

We have determined the strength of the force, $F_{2,1}$; we now have to determine the direction in which it acts. This is straightforward to determine if we make use of Fleming's left-hand rule (see Figure 8.13). We find that if the currents in the two

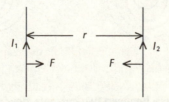

*Figure 8.16* **The magnetic force between two conductors**

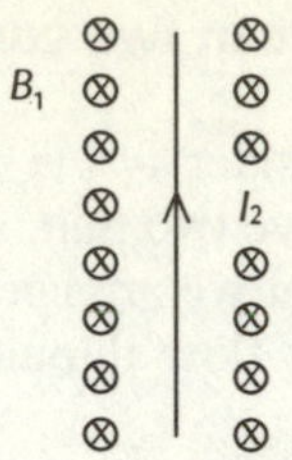

$$
\begin{matrix}
\otimes & \otimes \\
B_1 \ \otimes & \otimes \\
\otimes & \otimes \\
\otimes \quad \uparrow I_2 & \otimes \\
\otimes & \otimes \\
\otimes & \otimes \\
\otimes & \otimes \\
\otimes & \otimes \\
\end{matrix}
$$

*Figure 8.17* **The strength of the magnetic field due to a current flowing in the first conductor in the neighbourhood of the second conductor is $B_1$. ($\otimes$ indicates the magnetic field into the plane of the figure)**

conductors are flowing in the same direction, then the conductors attract each other. Figure 8.18(a) shows the magnetic fields due to each current separately. We note that the direction of the magnetic fields due to the two currents individually are in opposite directions in the region between the two conductors. The resultant magnetic field is shown in Figure 8.18(b). That the conductors attract each other is obvious from the properties attributed to the magnetic lines of flux.

It is straightforward to show that the force acting on the first conductor due to the current $I_2$ flowing in the second conductor, $F_{1,2}$, is also given by Equation (8.5).

*Special case*

Suppose we take the current flowing in each conductor, $I_1$ and $I_2$, to be one ampere, the conductors to be separated by one metre and the length of each conductor, $l$, to be one metre, then the force $F_{2,1}$ between the conductors is

$$
F_{2,1} = \frac{\mu_0}{2\pi} = 2 \times 10^{-7} \ \text{N}
$$

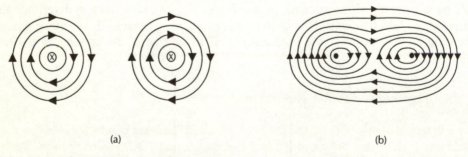

(a)                                        (b)

*Figure 8.18* **The magnetic force between two currents flowing in the same direction. ($\otimes$ denotes current flowing into the plane)**

■■■ **WORKED EXAMPLE 8.3**

Determine the magnitude and direction of the force on the middle conductor in Figure 8.19 if

$I_1 = 10$ A,   $I_2 = 10$ A   and   $I_3 = 10$ A

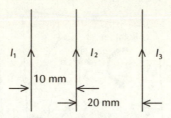

*Figure 8.19*   **The length of each conductor is 1 m**

– The force on the middle conductor due to the right-hand conductor is determined from Equation (8.5) as

$2 \times 10^{-4}$ N

and acts in a direction such that the conductors attract each other, whereas the force due to the left-hand conductor on the middle conductor is

$1 \times 10^{-4}$ N

Again, the direction of the force is such that the conductors attract each other.

The two forces which act on the middle conductor do so in opposite directions, and it follows that the net force acting on the middle conductor is

$F = (2 - 1) \times 10^{-4}$ N $= 1 \times 10^{-4}$ N

This observation leads on to the definition of the ampere:

■■■ **DEFINITION**

> The ampere is that current which, if flowing in two straight conductors of infinite length and of negligible cross-section, placed one metre apart in a vacuum, will produce a constant force of $2 \times 10^{-7}$ newton per metre of length on each of the conductors.

## 8.5   Introduction to the electric motor

As a practical application of the force on a current-carrying conductor in a magnetic field we describe the action of a simple direct current (DC) electric motor. Such a motor typically consists of a rectangular coil of wire mounted on an axle so that it can rotate between the poles of a permanent magnet; see Figure 8.20. Each end of

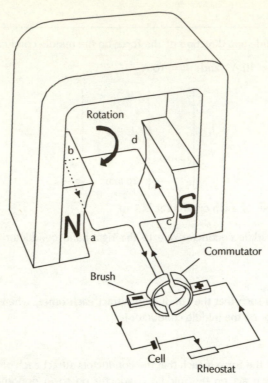

*Figure 8.20* **The simple DC electric motor**

the coil is connected to half a split conducting ring (the commutator) and are connected to a DC supply via two carbon blocks (the brushes) which are pressed lightly against the commutator.

Suppose the coil is in the horizontal position when the current is first switched on. Current will flow in the coil as indicated in Figure 8.20. Applying Fleming's left-hand rule, we note that the side ab of the coil experiences an upward force while the side cd a downward force; see Figure 8.21. These two forces are said to form a couple which causes the coil to rotate in a clockwise direction until it reaches the vertical position. In this position the brushes touch the space between the two halves of the commutator and the current ceases to flow. The coils inertia carries the coil past the vertical and the commutator halves change contact from one brush to the other. This reverses the current in the coil and, therefore, reverses the directions of the forces on the sides of the coil. The side ab is now on the right-hand side with a downward force on it, and side cd on the left-hand side with an upward force. The coil, therefore, continues to rotate in a clockwise direction for as long as the current flows.

Finally, we note that Fleming's left-hand rule is sometimes known as the motor rule since it was used to predict the direction of the force on the sides of the coil.

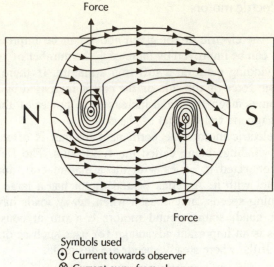

Force

N                                    S

Force

Symbols used
⊙ Current towards observer
⊗ Current away from observer

*Figure 8.21*   **The origin of the forces in the simple electric motor**

○ ▬▬ **WORKED EXAMPLE 8.4**

Determine the turning force or torque acting on the coil in Figure 8.20 if the magnetic field strength is $B$ and the current flowing through the coil is $I$. Take the sides ab and cd of the coil to have length $\ell$ and the side bc to have length $d$. Determine the numerical value of the torque if $\ell = 2.0$ cm, $d = 0.25$ cm and a current of 10 mA flows through the coil. Take the magnetic field to have strength 0.5 T.

– The magnitude of the force acting on each of the sides ab and cd is

$F = BI\ell$

and the forces act in opposite directions. These forces provide a turning force or torque, $T$, about the pivotal point or axle given as

torque = force × separation of forces

so that

$T = F \times d = BI\ell d$

Substituting the values for $B$, $I$, $\ell$ and $d$ we find the torque is

$T = 0.25 \ \mu\text{N m}$

Note that in Figure 8.20 the permanent magnet is shaped so that the magnetic field is always perpendicular to the plane of the coil.

## Practical electric motors

The description of the electric motor given above can be improved in many ways. This simple design can be improved by increasing the number of turns in the rotating coil, and also by winding them on a soft-iron armature. If there are *n* turns in the rotating coil then the force on the side of the coil is increased by a factor of *n*. The iron armature becomes magnetized and increases the power of the motor by adding the effects of its magnetic field to that of the coil.

In commercial electric motors the permanent magnet is often replaced with an electromagnet, its windings being called the field coils. The field coils are called *series wound* if connected in series with the armature coil, and *shunt wound* if connected in parallel with it. A series wound motor has a large turning effect, or torque, at low turning speeds, and is used when heavy loads have to be moved at rest. On the other hand, shunt wound motors can run at constant speeds under varying loads. This is an important advantage for uses such as driving power tools, such as lathes and drills, where steady speeds are desirable.

## 8.6 The magnetic effects due to moving electric charge

We have noted many times that an electric current is merely electric charge in motion. It follows that moving electric charges will both create magnetic fields (i.e. deflect a magnetic compass), and experience magnetic forces if they are moving in a magnetic field.

Consider an electric charge *q* moving with a speed *v* in a magnetic field, such that the charge cuts the lines of magnetic field, as indicated in Figure 8.22. Since an electric current is the flow of electric charge we expect a force to be acting upon the charge. The direction of force is given by Fleming's left-hand rule and is shown in the figure. Note that this assumes that the charge *q* is positive (recall that conventional current flows in the direction in which positive charge moves). If the charge *q* were actually negative, then the conventional current associated with the charge would be in the opposite direction to the charge movement. It follows that the force acting upon the charge would be in the opposite direction to that indicated in Figure 8.22.

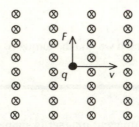

*Figure 8.22* **The magnetic force acting on a moving electric charge. (⊗ indicates the magnetic field into the plane of the figure)**

The magnitude of the force is given by the equation

$$F = Bqv \tag{8.6}$$

Rather than derive this, we show that the right-hand side of (8.6) has the correct units. This is straightforward if we compare (8.6) with the result for the force acting on a current flow (Equation (8.3)), and note that $qv$ has the units of

$$C\,\frac{m}{s}$$

while $Il$ has the units of

$$\frac{C}{s}\,m$$

which are identical. Hence the right-hand side of (8.6) has the same units as the right-hand side of (8.3), i.e. the result (8.6) has the units of force. We note that this force is always perpendicular to the motion of the charge. It follows that the resulting path of the charge is that of a circular motion.

As an interesting example consider Worked Example 8.5.

━━━ **WORKED EXAMPLE 8.5**

The electron is found to behave as if it is spinning; see Figure 8.23:

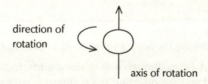

direction of
rotation

axis of rotation

*Figure 8.23*

— A rotating charge implies that charge is in motion, i.e. that an electric current is flowing; this, in turn, suggests that the electron should have a magnetic field associated with it.

Finally, we note that stationary charges do not experience any magnetic forces when placed in a magnetic field – this follows because if the charges are stationary, then there is no associated electric current and, therefore, no associated magnetic field.

## 8.7 Summary

We have seen that there is a connection between electricity and magnetism. In particular, we pointed out that a conductor carrying an electric current exerts a force on a permanent magnet. Similarly, a magnetic field was found to exert a force on a conductor carrying an electric current. We explained these results by arguing that an electric current (i.e. electric charge in motion) creates a magnetic field.

Note that we did not explain *how* or *why* an electric current creates a magnetic field. All we know is that when a compass is placed near a conductor carrying a current it is deflected. Since the compass responds to magnetic fields, we must have that the current creates a magnetic field. There is little else we can suggest as an explanation.

Given that there is a magnetic field associated with an electric current we went on to discuss two properties of this field that interested us, namely, the pattern of the field and what the strength of the field depends upon.

The pattern of the magnetic field depends on the path the current takes. It is found that the simplest pattern is for a long straight wire, where the lines of magnetic flux were found to take the form of concentric circles.

It follows that, since a current creates a magnetic field then, if a current-carrying conductor is placed in the region of a magnetic field, there will be magnetic forces exerted on it. This led on to an introductory discussion of the electric motor for which an expression for the turning moment or torque was derived. Some practical points of electric motors were discussed.

Finally, it was noted that since an electric current is merely electric charge in motion, it follows that moving electric charges will both create magnetic fields (i.e. deflect a magnetic compass) and, therefore, also experience magnetic forces if they are moving in a magnetic field.

## Problems

In the following problems take the numerical value of $\mu_0$ in SI units to be $\mu_0 = 4\pi \times 10^{-7}$.

**8.1** A current of 250 mA flows through a long straight conductor. Determine the strength of the magnetic field and the direction the North pole of a small compass would point in if placed at points X, Y, and Z in Figure P8.1.

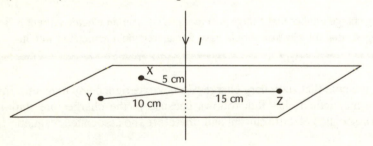

*Figure P8.1*

**8.2** Determine the force, i.e. magnitude and direction, on the conductor placed in each of the magnetic fields shown in Figures P8.2(a) and (b).

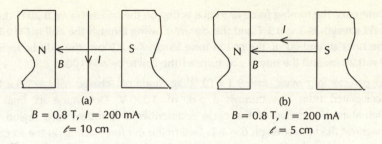

(a)

$B = 0.8$ T, $I = 200$ mA

$\ell = 10$ cm

(b)

$B = 0.8$ T, $I = 200$ mA

$\ell = 5$ cm

*Figure P8.2*

**8.3** Determine the force, i.e. magnitude and direction, on each conductor shown in Figure P8.3.

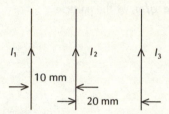

*Figure P8.3* **Three conductors (each 1 m long)**

(a) $I_1 = 10$ A, $I_2 = 20$ A, $I_3 = -10$ A, (b) $I_1 = 10$ A, $I_2 = 10$ A, $I_3 = 20$ A.

**8.4** Figure P8.4 shows a conductor of radius 200 mm situated in a radial field provided by a pair of shaped pole pieces. Determine the magnitude and direction of the force acting on the conductor if a current of 2 A flows through it.

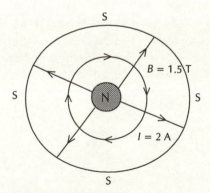

*Figure P8.4*

*Hint:* Consider a small element of the conductor and determine the direction of the force acting on it. Show that the force acting on all small elements are in the same direction. Therefore, the result (6.3) is valid here, with $l$ taken as the total length of the conductor.

o **8.5**   Determine the turning force or torque acting on the coil in Figure 8.20 if the magnetic field strength is $B = 0.5$ T and the current flowing through the coil is $I = 20$ mA. Take the sides ab and cd of the coil to have length $\ell = 2.5$ cm, the side bc to have length $d = 0.20$ cm and the number of turns of the coil to be $n = 100$.

o **8.6**   A particle of mass $m = 9.1 \times 10^{-31}$ kg and of charge of $-1.6 \times 10^{-19}$ C is accelerated from rest through a p.d. of 1000 V. Determine its final speed, $v$. Determine the radius of the circle it describes if it moves to a region where the magnetic field has strength $B = 2$ T. Determine the force acting on the electron due to the magnetic field. Show by direct calculation that this is equal to the centripetal force $mv^2/r$.

o **8.7**   A particle of charge $q$ and mass $m$ moves with a speed $v = 10^7$ m s$^{-1}$ and describes a circle of radius $r = 0.0284$ mm in a magnetic field of strength $B = 2$ T. Determine the charge to mass ratio, i.e. $q/m$, of the particle.

# CHAPTER NINE

# *Electromagnetic induction*

We have seen that there is some connection between electricity and magnetism. In particular, we pointed out that a conductor carrying an electric current exerts a force on a permanent magnet. Similarly, a magnetic field was found to exert a force on a conductor carrying an electric current. We explained these results by arguing that an electric current (i.e. electric charge in motion) creates a magnetic field. In this and the next chapter we develop this link further.

In these cases we were looking at the effect of *moving* charges on *steady* or *stationary* magnetic fields. It is of interest to determine if there are any effects due to *changing* magnetic fields on *stationary* charges. We will find that this will lead us on to the concept of

electromagnetic induction

By considering simple experiments we will be led to

Faraday's law

which gives the size of the induced e.m.f. when a conductor incurs a change of magnetic flux. We will be interested in the direction or the polarity of this induced e.m.f.. Related to the polarity of the induced e.m.f. is the direction of the accompanying induced current flows. We will find this is easily predicted by

Fleming's right-hand rule

A different approach to this is given by

Lenz's law

which has as its basis the Law of Conservation of Energy.

Finally, we will discuss a practical application of electromagnetic induction, namely the generation of an e.m.f. due to a coil rotating in a magnetic field.

## 9.1 Introduction to electromagnetic induction

Figure 9.1 shows a conductor placed in the uniform magnetic field supplied by a shaped permanent magnet. The magnet is shaped to give a uniform magnetic field between the two poles. This simplifies the experimental analysis. The conductor is connected to an ammeter. Experimentally, we find that if the conductor is stationary then the ammeter registers that there is no electric current. Certainly, there is no particular reason to expect a current in such a situation.

Let us consider the situation when the conductor moves in such a manner that it 'cuts' the lines of magnetic flux, as shown in Figure 9.2. We also indicate in the figure that the conductor is made up of equal amounts of positive and negative charge. From previous results (see Section 3.1, in particular Figure 3.7 and the related discussion), we know that no electric current flows when the conductor moves. This follows since the conductor is overall electrically neutral. An important point to note, however, is that separately both positive charge and negative charge are in motion and it is the cancellation of the electric currents associated with the movement of both charges that results in no overall current flow.

Should we expect anything if the conductor moves through a magnetic field? Let us consider the magnetic forces acting on both the moving positive charge and the moving negative charge separately. Making use of Fleming's left-hand rule (see Figure 8.13), we find that the direction of the force acting on the positive charge is towards the point Y. Similarly, the force acting on the negative charge is towards the point X (we can either make use of Fleming's left-hand rule again, remembering the importance of conventional current flow, or merely note that the magnetic force on a negative charge will be in the opposite direction to that of a positive charge moving in the same direction).

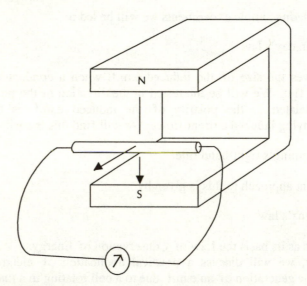

*Figure 9.1* **Electromagnetic induction**

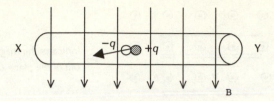

*Figure 9.2*   **Movement of a conductor through a magnetic field**

Thus, although there is no net current flow in the conductor, the magnetic force acting on the charges making the conductor are such that there is a charge separation. Suppose in Figure 9.2 we connect an ammeter to the conductor, thus completing the circuit. The effect of this is to provide a path for both the positive and negative charges to move completely around the circuit. Thus, the positive charge is pushed through the ammeter, around the circuit, in one direction, while the negative charge is pushed through the ammeter and around the circuit in the other direction. Hence, the ammeter will register a flow of electric current; this current is referred to as the

     induced current

We point out that, given we have previously discussed the magnetic forces which act on electric currents, i.e. electric charge in motion, this phenomena, referred to as *electromagnetic induction* should not really be too much of a surprise, and indeed it may even have been expected.

Suppose the conductor is moved in a direction along the lines of magnetic flux. Do we expect a current to flow in this situation? The arguments leading to Equation (8.4) suggest that we should expect no forces to act on either the positive or negative charges that make up the conductor. If no forces are acting on the charges making up the conductor, there will be no charge separation and, therefore, no induced current will be expected to flow in a complete circuit.

Note that it is not merely the presence of the magnetic field which produces the current, but also the motion of the conductor through the field. Thus, to generate an induced current, it is important for the wire to 'cut' the lines of magnetic flux as indicated in Figure 9.2.

We now wish to discuss the factors that determine the size of the induced current. Careful experiments show that the induced current depends upon the product of the following three factors:

- the strength of the external magnetic field, $B$,
- the length of conductor in the magnetic field, $l$, and
- the speed with which the conductor crosses the magnetic field, $v$,

i.e.

     induced current $\propto Blv$                                                    (9.1)

Is the result (9.1) reasonable? We first note that, since we have explained the phenomena in terms of the action of magnetic forces, that the induced current

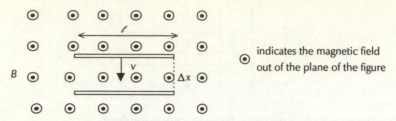

*Figure 9.3*   **The motion of the conductor through a magnetic field**

depends upon the external magnetic field $B$ is expected, if the magnetic field increases, the forces acting on the moving positive and negative charges increases. When charge is in motion, the current associated with the charge depends upon the speed $v$, so that (9.1) depends upon the speed of the conductor is also reasonable. Finally, if the length of the conductor in the magnetic field, $l$, increases, the number of positive–negative charge pairs affected increases. If more charge is displaced, the electric current increases.

To proceed further we have to consider the expression for the induced current and write in it a more general form. We begin by noting that if the speed of the conductor is $v$, in a time $\Delta t$, the conductor has moved a distance $\Delta x$ given as

$$\Delta x = v \, \Delta t$$

See Figure 9.3.

From this result, we have that the speed $v$ is related to the distance $\Delta x$ moved in the time $\Delta t$ as

$$v = \frac{\Delta x}{\Delta t}$$

and, thus, the expression $Blv$ can be written as

$$Blv = Bl \, \frac{\Delta x}{\Delta t}$$

Before proceeding, and interpreting the above result, we first have to introduce the concept of magnetic flux or magnetic flux linkage; see Figure 9.4.

We have previously indicated that the strength of the magnetic field could be thought of in terms of the density of the lines of magnetic flux (see Section 8.1 and the discussion relating to Figure 8.2 in particular). Indeed, we have said that the magnetic strength, $B$, is referred to as the 'magnetic flux density'. Closely related to this is the amount of magnetic flux, $\Phi$, linking a circuit; see Figure 9.4.

With reference to Figure 9.4, suppose we have a uniform magnetic field $B$ and we

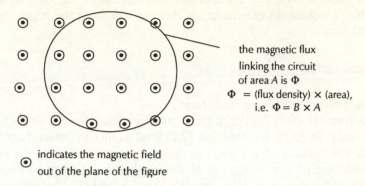

the magnetic flux
linking the circuit
of area $A$ is $\Phi$

$\Phi$ = (flux density) × (area),
 i.e. $\Phi = B \times A$

⊙ indicates the magnetic field
out of the plane of the figure

*Figure 9.4* **The magnetic flux linking a circuit**

place a circuit of area $A$ normal to the field, i.e. so that the lines of magnetic flux are perpendicular to the area $A$. Here, the term 'circuit' is meant to indicate a conductor which has its ends joined together. We define the amount of magnetic flux linking the circuit as

$$\Phi = B \times A$$

The terminology is reasonable since we have a quantity related to a density, with units of $Wb\,m^{-2}$, multiplied by an area, with units of $m^2$, to give a quantity with units of Wb, i.e.

$$Wb\,m^{-2} \times m^2 = Wb$$

Since we interpreted 'magnetic flux density', $B$, as lines of magnetic flux per unit area, we can interpret the 'magnetic flux', $\Phi$, as the total number of magnetic lines of flux.

We now return to our discussion, noting that $l\Delta x$ is just the area swept out by the conductor, call it $\Delta A$, in the time interval $\Delta t$. It is convenient to call the quantity $B \times \Delta A$ the *change* in the flux linkage, $\Delta\Phi$, linking the circuit

$$\Delta\Phi = B \times \Delta A$$

The quantity $\Delta\Phi$ is the amount of the magnetic field which is 'cut' by the conductor as it moves, and is also referred to as the *change* in the magnetic flux. It can be thought of as the number of lines of magnetic field which are cut as the conductor sweeps out its path. Alternatively, it can be thought of as the change in the number of magnetic lines of flux which links the area $\Delta A$.

Notice that, since the SI unit of the magnetic field strength, $B$, is the tesla (T) or the weber per square metre ($Wb\,m^{-2}$), suitable SI units for the flux linkage are the tesla square metre ($T\,m^2$) and the weber (Wb). We will generally refer to the unit of flux linkage and change in flux linkage as the weber.

Thus, we have shown that the quantity *Blv* measures the rate at which the magnetic flux is cut by the conductor, so that we can write the result (9.1) in the more general form

$$\text{induced current} \propto \frac{\Delta\Phi}{\Delta t} \tag{9.2}$$

Two important comments can be made here:

*(1)* The form of (9.2) for the induced current given above is considered to be more important than (9.1). We note that (9.2) implies that an induced current flows in the conductor because of a change in the magnetic flux linking the conductor, rather than because of any movement of the conductor. This can be demonstrated by making a varying current flow through, say, a coil. If the current flowing through the coil varies, the magnetic field created by the current also varies. If a conductor is placed close by, the magnetic field linking the conductor will vary. This means that there will be a change in the magnetic flux linking the conductor (see Figure 9.5). In this situation it would be difficult to see how we are to make use of (9.1), since no physical movement is involved. In principle, the use of (9.2) is straightforward; we merely have to calculate the change in the magnetic flux linking the conductor and then divide by the time $\Delta t$ in which the change occurs in (in practice the calculation for the situation envisaged in Figure 9.5 may be quite difficult).

Note that in this situation, we have a stationary conductor and a changing magnetic field, a relationship we suggested we wished to consider at the beginning of this chapter. The transformer is a device which makes use of this phenomenon and is described in greater detail in Section 10.4.

*(2)* We have emphasized the induced current in the conductor. For a current to exist there must be an 'e.m.f.-like' source in the circuit. Unlike a source of e.m.f. such as, say, a battery, the *induced e.m.f.* is not located or local to one part of the circuit. The current only exists when there is a complete circuit, but the e.m.f., which drives the current, is present even if the circuit is broken. Since the amount of current for a given e.m.f. depends upon the total resistance presented by the circuit to the flow of charge, we might expect a simpler relationship between the induced

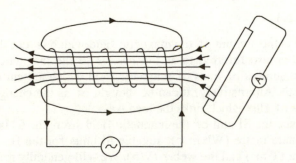

*Figure 9.5* **Generation of an e.m.f. due to a changing magnetic field and a stationary conductor**

e.m.f. and the flux change and write

$$\mathcal{E} \propto \frac{\Delta\Phi}{\Delta t} \tag{9.3}$$

Note, we use the symbol $\mathcal{E}$ for the induced e.m.f. to distinguish it from the e.m.f. of a source such as a battery, $E$. If, in an experiment, the change in the magnetic flux linking the conductor is $\Delta\Phi$ and is measured in webers, and the time $\Delta t$ for the change is measured in seconds, then the induced e.m.f. $\mathcal{E}$ is given directly in volts and we can write

$$\mathcal{E} = \frac{\Delta\Phi}{\Delta t} \tag{9.4}$$

This is referred to as Faraday's law.
Worked Example 9.1 shows a straightforward use of (9.4).

■■■■ **WORKED EXAMPLE 9.1**

A conductor of length 20 cm moves through a region where there is a constant magnetic field of strength 2 tesla with a speed of $3 \text{ ms}^{-1}$ (see Figure 9.6). Determine the e.m.f. induced across the conductor making use of (9.4). If the conductor has a resistance of $2 \Omega$, what current flows through the conductor?

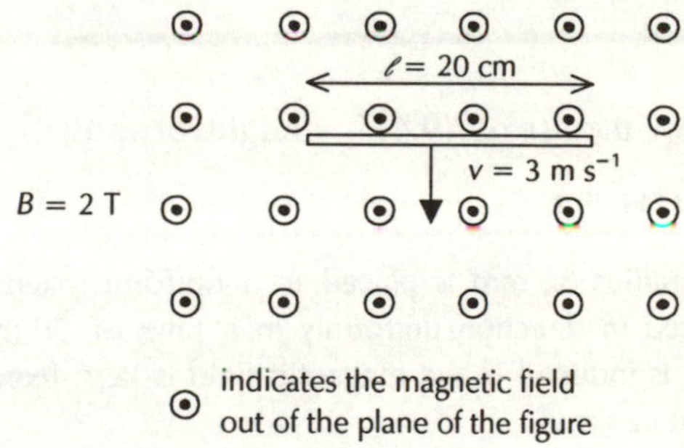

indicates the magnetic field
out of the plane of the figure

*Figure 9.6*

– In each second the conductor sweeps out an area of

$$\ell(v \times 1 \text{ s})$$

and so encloses an amount of magnetic flux

$$\Delta\Phi = B\ell(v \times 1 \text{ s}) = 1.2 \text{ Wb}$$

The e.m.f. induced, $\mathcal{E}$, is given by (9.4) with $\Delta t = 1$ s. We find

$$\mathcal{E} = \frac{\Delta\Phi}{\Delta t} = \frac{1.2 \text{ Wb}}{1 \text{ s}} = 1.2 \text{ V}$$

Note that there is no current in the conductor since there is no complete circuit.

This example effectively shows that the e.m.f. generated between the ends of the conductor when moving through a magnetic field in the manner suggested by Figure 9.6 is given by the result

$$\mathscr{E} = Blv \tag{9.5}$$

Worked Example 9.2 makes use of the result (9.5) directly, and describes an experiment attempted during a flight of the Space Shuttle. The experiment was known as the so-called 'tether experiment' and took place during August 1992.

■■■■ **WORKED EXAMPLE 9.2**

What is the e.m.f. generated between the ends of a conductor 20 km long in the magnetic field in a near-Earth orbit?
− As has been mentioned previously, the Earth has a magnetic field. If a conductor moves through this field it follows that an e.m.f. will be generated, assuming that the conductor cuts the lines of magnetic flux. To determine the size of the induced e.m.f. we need to know both the speed at which the conductor is moving through the magnetic field and the value of the strength of the magnetic field. Assuming that the experiment took place in a circular near-Earth orbit, the speed of the shuttle will be approximately 8 km s$^{-1}$ and the strength of the magnetic field has a value of approximately $5 \times 10^{-5}$ T. Making use of (7.5), the induced e.m.f. is given as

$$\mathscr{E} = B\ell v \approx 8000 \text{ V}$$

The final example of the use of (9.4) is straightforward:

■■■■ **WORKED EXAMPLE 9.3**

A coil of 50 turns of radius 30 mm is placed in a uniform magnetic field of 2 T. If the magnetic field is reversed in direction uniformly in a time of 50 ms, determine the e.m.f. generated. What e.m.f. is induced if the magnetic field is kept fixed and the coil is flipped over in the same amount of time?

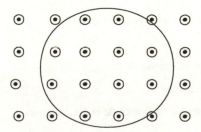

⊙ indicates the magnetic field out of the plane of the figure

*Figure 9.7*

For one turn the initial magnetic flux linkage is

$$\Phi = B \times A = (2\text{ T}) \times (0.002\,827\text{ m}^2) = 5.655 \times 10^{-3}\text{ Wb}$$

Hence, for 50 turns the initial magnetic flux linking the circuit is

$$\Phi = 50 \times 5.655 \times 10^{-3}\text{ Wb} = 2.283 \times 10^{-1}\text{ Wb}$$

When the magnetic field reverses direction there will be a change in the magnetic flux linking the circuit so that an e.m.f. will be induced. The change in flux linking the coil is

$$\Delta\Phi = 2 \times 2.283 \times 10^{-1}\text{ Wb} = 5.655 \times 10^{-1}\text{ Wb}$$

The e.m.f. generated is given by (9.4) with $\Delta t = 50$ ms,

$$\mathscr{E} = \frac{\Delta\Phi}{\Delta t} = \frac{5.655 \times 10^{-1}\text{ Wb}}{50 \times 10^{-3}\text{ s}} = 11.310\text{ V}$$

If the coil is flipped over in the same time and the field is kept fixed, the change in the flux is the same as that calculated above, so that the e.m.f. induced in the coil is the same.

---

### Fleming's right-hand rule

Besides being interested in the size of the induced e.m.f., we are also interested in the *polarity* or sense of the induced e.m.f. This determines the direction in which the induced current flows (assuming a complete circuit). A convenient rule for determining the direction of the induced current, given the direction of the magnetic field and the motion of the conductor, is given by Fleming's right-hand rule, see Figure 9.8.

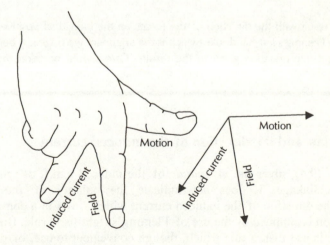

*Figure 9.8* **Fleming's right-hand rule**

*Table 9.1*    **Fleming's right-hand rule**

| | |
|---|---|
| First finger | points in the direction of the magnetic **F**ield |
| Se**C**ond finger | points in the direction of the induced **C**urrent |
| Thu**M**b | points in the direction of the **M**otion of the conductor |

When using this rule, the first finger, second finger and thumb of the right hand are placed in mutually perpendicular positions, as indicated in the figure. The first finger is taken to point in the direction of the magnetic field, while the thumb points in the direction that the motion of the conductor. The second finger points in the direction of the induced current flowing in the conductor. These results are collected in Table 9.1.

### ■ WORKED EXAMPLE 9.4

Applying Fleming's right-hand rule to Figure 9.9(a), the direction of the induced current, $I$, is as shown. If we were to think in terms of a source of e.m.f., such as a battery, the polarity of the source would have to be chosen as in Figure 9.9(b).

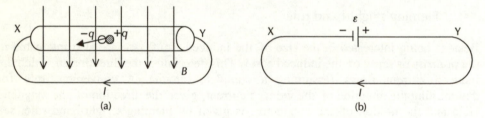

*Figure 9.9*    **(a) Application of Fleming's right-hand rule; (b) the polarity of the induced e.m.f.**

Compare this result with the direction of the forces on the individual positive and negative charges given by Fleming's *left*-hand rule (which is the argument we originally began with), but note the different interpretation given to the results. Care should be taken not to confuse them.

## 9.2 Lenz's law and the direction of the induced current

Faraday's law, (9.4), gives the *magnitude* of the induced e.m.f. due to a changing magnetic flux linkage. It does not indicate the polarity of the e.m.f., or, equivalently, the direction of the induced current which flows in a complete circuit. This information is supplied by the use of Fleming's right-hand rule. But, Fleming's right-hand rule is just that, a rule which, though convenient to use, offers no reason why it works. Lenz provided another approach to this problem and gave a convincing explanation of why an induced current should flow in a particular direction. Lenz formulated his approach in the form of a law which states:

> The direction of the induced e.m.f. is always such as to oppose the change producing it.

Figures 9.10 (a) and (b) illustrate the application of this law in the case of a magnet moving with respect to a coil.

When the S pole of the magnet is moved towards the coil, the amount of magnetic flux linking the circuit is increasing. According to Lenz's law the current

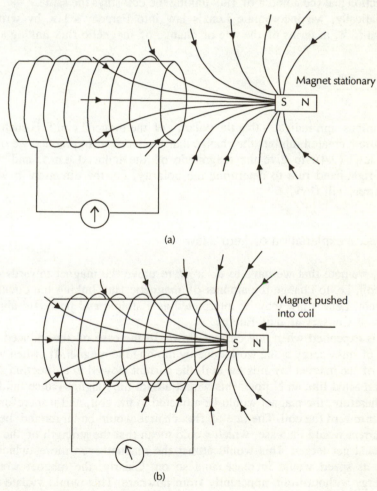

(a)

(b)

*Figure 9.10*   **An application of Lenz's law**

induced must flow in a direction so as to give an S polarity to the end of the coil facing the magnet. Why should this be? If the induced current flows in a direction such that an S pole is induced, then, since like poles repel, the magnet feels a repulsive force which tends to oppose further motion. Lenz's law states the induced current will flow in the coil in such a direction so as to oppose the change. This means the induced current must flow in that direction such that the amount of flux linking the coil stays the same. The only way this can happen is if the induced current flows so as to try and repel the magnet moving towards the coil.

Similarly, when the S pole of the magnet is pulled away from the coil the direction of the induced current is reversed. The near end of the coil now acts like an N pole and, since unlike poles attract, the motion of the magnet is again opposed. Here, the magnetic flux linking the coil is getting smaller, so that the induced current flows in such a direction that the amount of flux linking the coil stays the same.

Mathematically, we incorporate Lenz's law into Faraday's law by writing the induced e.m.f., $\mathscr{E}$, in terms of the rate of change of magnetic flux linking a circuit, $\Delta\Phi/\Delta t$, as

$$\mathscr{E} = -\frac{\Delta\Phi}{\Delta t} \tag{9.6}$$

where the minus sign indicates that the polarity of the induced e.m.f. is such that the induced current created inhibits the change that produced it. In practice we often use Faraday's law, (9.4), to give the magnitude of the induced e.m.f. and then use Fleming's right-hand rule to determine the polarity, i.e. the direction in which an induced current will flow.

○      Physical explanation of Lenz's law

In general, we note that we have to *do work* to move the magnet towards or away from the coil, i.e. to change the amount of magnetic flux linking the circuit. From this comment, Lenz's law may be interpreted as a simple and powerful application of the Law of Conservation of Energy.

Energy is expended when any current flows. In the case of the induced current, the source of this energy is the work done by moving the magnet. If, when we move the S pole of the magnet towards the coil, the current flowed in a direction such that the near end acted like an N pole, this would attract the magnet (since unlike poles attract). Therefore, the magnet would be attracted to the coil, and it therefore would move faster toward the coil. The rate of flux change would be larger and, hence, the induced current would increase, which would mean that the strength of the induced N pole would get larger. This would attract the magnet even more, which would mean that its speed would increase, and so on. Clearly, the magnet would gain kinetic energy without limit, apparently from nowhere. This would violate the Law of Conservation of Energy and is therefore forbidden.

It follows, therefore, the induced current must flow in the direction such that the end of the coil acts like an S pole.

## 9.3 The generation of an e.m.f. due to a coil rotating in a magnetic field

As a practical application of electromagnetic induction we consider the generation of an e.m.f. due to a coil rotating in a constant magnetic field. According to previous results, when a conductor moves such as to cut the lines of magnetic flux of a magnet, an e.m.f. is induced between the ends of the conductor. The magnitude of the e.m.f. is proportional to the rate of cutting of the lines of flux; recall the discussion in Section 9.1 in general, and Faraday's law (9.4) in particular.

Figure 9.11 depicts a simple arrangement for the continuous production of an e.m.f. which we will spend some time discussing. The figure shows a rectangular coil of wire which rotates uniformly in the magnetic field between the poles of a permanent magnet. The ends of the coil are connected to two *slip rings* mounted on the coil spindle. Current may be obtained from the coil through two *brushes*, which are made to press lightly against the slip rings.

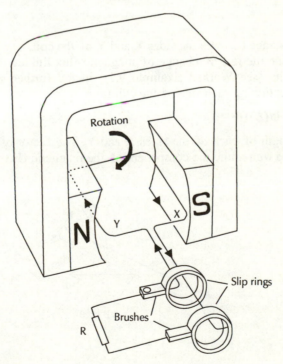

*Figure 9.11*  **The simple electric generator**

As the coil rotates, its sides cut the magnetic flux and therefore an e.m.f. is induced. Because there is a complete circuit an induced current can flow through the circuit. For the particular coil movement indicated in the figure, i.e. with the side X moving instantaneously downwards and side Y moving instantaneously upwards, an application of Fleming's right-hand rule suggests that the induced current flows along the side X via the slip rings, then back along the side Y. Note that when viewed from above the current flows in a clockwise direction through the coil.

We make use of Figure 9.12 to consider the variation of the induced e.m.f. with time over one complete rotation. The coil is taken to be rotating in a uniform magnetic field with a constant angular speed or angular frequency $\omega$. The magnet is shaped such that the coil rotates in a constant magnetic field of strength $B$, independent of the orientation of the coil in the field. Suppose we take the time $t$ equal to zero to be the instant when the coil is in the vertical position with its side X uppermost. After a further time $t$, the coil has rotated through the angle $\theta$ with respect to the magnetic field, which is given simply as

$$\theta = \omega t$$

where $\theta$ is measured in radians.

Note that the speed of each end of the coil, $v$, is constant and is given as

$$v = \omega \frac{b}{2}$$

where $b$ is the distance between the sides X and Y of the coil.

Let us determine the rate of change of magnetic flux linking the coil. According to previous results (see Worked Example 9.1), in the further small time $\Delta t$, the change in the flux linking each side of the coil is

$$\Delta\Phi = (B \sin(\theta))lv$$

where $l$ is the length of each of the sides X and Y. The factor of $\sin(\theta)$ appears in the above because we require the component of the magnetic flux which is being cut

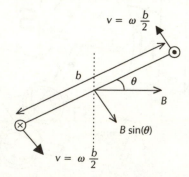

*Figure 9.12*   **The coil moving through a uniform magnetic field**

by the coil. Hence, it follows that $\mathcal{E}$, the e.m.f. induced across the ends of each side of the coil, is given as

$$\mathcal{E} = \frac{\Delta\Phi}{\Delta t} = 2Blv\sin(\theta)$$

The factor of 2 appears in the above because, although the polarities of the e.m.f. induced on each side of the coil are in opposite directions, both cause the induced current to flow in the same direction. It therefore follows that the e.m.f. generated across both ends of the coil is given as twice that for each side considered alone.

Substituting for the angle $\theta$ in terms of $\omega$ and $t$ and for $v$ in terms of $\omega$ and $b$, we find that the induced e.m.f., $\mathcal{E}$, is given as

$$\mathcal{E}(t) = Bl\omega b\sin(\omega t) \tag{9.7}$$

One obvious point to note is that the induced e.m.f. depends upon time, i.e. as the time varies so does the magnitude and polarity of the e.m.f. We emphasize this dependence by writing $\mathcal{E}$ as $\mathcal{E}(t)$.

Figure 9.13 shows how the induced e.m.f. varies over one complete rotation. Let us now check if this result is what we expect. We take the starting point to be the instant that the coil is in the vertical position with its side X uppermost (see Figure 9.13(a)).

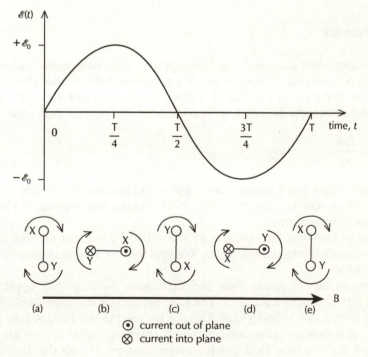

Figure 9.13 **The induced e.m.f. over one complete cycle**

In this position the sides of the coil are moving along the lines of magnetic flux and, therefore, the induced e.m.f. at this instant is zero. This follows since no lines of magnetic flux are being cut in this position. During the first quarter of the rotation the induced e.m.f. increases from zero to its maximum or *peak value* when the coil is in the horizontal position (see Figure 9.13(b)). For the next quarter, the induced e.m.f. decreases as the component of the magnetic field perpendicular to the coil becomes smaller and again reaches zero when the coil is in the vertical position with the side Y uppermost (see Figure 9.13(c)). During the second half of the rotation the induced e.m.f. follows the same pattern as described above, except that the polarity of the induced e.m.f. is reversed. This follows because the role of the sides X and Y across the magnetic flux are reversed.

This type of supply is referred to as a sinusoidal voltage supply. If the supply was connected across a resistor we would find that the current would vary in the same manner as the e.m.f. For this reason the supply is referred to as an alternating current or AC supply. The term AC indicates that the current changes direction with time. This distinguishes it from a direct current or DC supply, such as a battery, where the e.m.f. and the current do not change direction with time.

Note that we made use of Fleming's right-hand rule to determine the polarity of the induced e.m.f., so that sometimes the rule is known as the 'dynamo rule', dynamo being another term for a generator.

### 9.4  Summary

In this chapter we discussed the concept of electromagnetic induction, first considering a conductor moving through a magnetic field. It was then pointed out that the important concept was that of a change in the magnetic flux linking a circuit. This enabled us to write the e.m.f. induced between the ends of the conductor as

$$\mathscr{E} = \frac{\Delta \Phi}{\Delta t}$$

This formula, known as Faraday's law, applies to situations where, for example, the conductor is at rest but the magnetic field linking the conductor changes, for example, as in a transformer.

We want to know not only the size of the induced e.m.f. but also the direction in which the accompanying current flows. We find this is easily predicted by Fleming's right-hand rule.

It is convenient to stress here the difference between the applications of Fleming's left- and right-hand rules. The rules are similar and care should be taken to ensure the correct rule is being used. Usually we use the left-hand rule to predict the motion of a current carrying conductor in a magnetic field, i.e. we are given the direction of the magnetic field and the current and we predict the direction of the force. The right-hand rule is used to predict the direction of an induced current due to the motion of a conductor in a magnetic field, i.e. we are given the direction of the magnetic field and the motion of the conductor current and we predict the direction of the induced current.

We then presented a brief discussion of Lenz's law, which provided a more rigorous explanation for the direction in which the induced current flows. We showed that Lenz's law may be interpreted as a particular case of the more general law of Conservation of Energy.

We finished with a discussion of a practical application of electromagnetic induction, namely, the generation of an e.m.f. due to a coil rotating in a magnetic field.

## Problems

**9.1**    Consider a conductor of length $\ell = 2$ m moving with a speed $v = 0.5$ m s$^{-1}$ in a uniform magnetic field $B = 3$ T, as indicated in Figure P9.1.

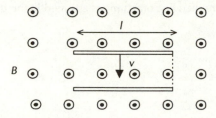

indicates the magnetic field out of the plane of the figure

*Figure P9.1*

    (a)  Determine the e.m.f. induced in the conductor.

    (b)  If the circuit is complete, indicate the direction in which the current flows, quoting the relevant rule.

**9.2**    A conductor of radius $r = 25$ cm is placed in a uniform magnetic field of strength $B = 2.5$ T.

    (a)  Determine the flux, $\Phi$, linking the circuit.

    (b)  If the magnetic field is reversed determine the change in flux, $\Delta\Phi$, linking the circuit.

    (c)  If this change takes place uniformly in a time $\Delta t = 5$ ms, determine the e.m.f. induced in the coil.

    (d)  Determine the induced e.m.f. if the magnetic field is kept constant and the coil is flipped in the same time. Explain your result.

**9.3**    A coil of radius 50 cm is composed of 100 turns. The coil is placed in a uniform magnetic field of strength $B$. If the induced e.m.f. in the coil is 20 V due to the magnetic field uniformly falling to zero in 20 ms, determine the initial strength of the magnetic field.

**9.4**    It is arranged that the magnetic field linking a coil of area 10 cm$^2$ and 50 turns depends upon time in the manner as shown in Figure P9.2. Determine the e.m.f. induced in the coil during the times AB, CD and EF.

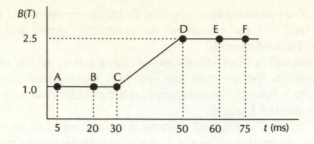

*Figure P9.2*

**9.5** An AC generator consists of a coil consisting of 100 turns of wire rotating with an angular speed of $\omega = 314 \text{ rad s}^{-1}$ in a uniform magnetic field of $B = 1.5$ T. If the area of the coil is 200 mm² determine the maximum and minimum values of the e.m.f. induced in the coil.

*Hint:* Make use of (9.7) and note that we are given the area of the coil. Also take into account that the coil consists of more than just one turn, as assumed in the derivation of (9.7).

# CHAPTER TEN

# *Inductance*

We have seen in the last two chapters that there is some connection between electricity and magnetism. In particular, we noted that a conductor carrying an electric current exerts a force on a permanent magnet. Similarly, a magnetic field was found to exert a force on a conductor carrying an electric current. We explained these results by arguing that an electric current creates a magnetic field. It was pointed out that these results were to do with the interaction between moving charges and steady magnetic fields.

We then considered if we might expect any effects between stationary charges and changing magnetic fields. We found that changing magnetic fields do influence stationary charges, causing them to move (i.e. a current to flow); this effect is referred to as electromagnetic induction.

We now develop the subject of electromagnetic induction a bit further and discuss the concept of *self-inductance* or, more simply, *inductance*.

## 10.1  Introduction to inductance

We have pointed out that a changing magnetic field induces a current through a conductor (more exactly, an induced e.m.f. across the ends of the conductor, a current only flows if there is a complete circuit). An important point is that this induced current flows in a direction which tends to oppose the change which produced it (see Figure 9.10). With the magnet and the coil stationary with respect to each other, the ammeter registers no current flowing in the circuit. If we push the S pole of the magnet towards the coil of wire, the magnetic field linking the circuit increases compared with its previous value.

Technically, we say the magnetic flux linking the circuit, or the flux linkage, has changed. The current induced in the coil flows in such a direction as to reduce this increase, that is, to make the magnetic field linking the circuit the same as its previous value. It was pointed out that as the magnet is brought closer to the coil, the induced current flows in a direction such that we have to expend energy.

We now want to think of the above in terms of a property of the coil referred to as its

### self-inductance

The term self-inductance is more simply referred to as inductance. We can conveniently discuss this with reference to Figure 10.1 and by considering the behaviour of a long coil when we try to change the current flowing through it. We will note the importance of both Faraday's law and Lenz's law in its behaviour.

The pattern of the magnetic field inside a long coil was discussed in Section 8.2. We found that inside a long coil of length $l$ and with $n$ turns, the magnetic field, $B$, created when a current $I$ flows through the coil was of uniform strength and direction, with a value given by

$$B = \frac{\mu_0 n I}{l}$$

(see Figure 8.10 and Equation (8.2)).

If the coil has radius $r$, the magnetic flux linking the circuit, $\Phi$, contained in or linking each turn of the coil, is given as

$$\Phi = B \times A = B\pi r^2$$

since the magnetic flux, 'number of lines of flux', is just the magnetic flux density multiplied by the area (recall Figure 9.4 and the related discussion). It follows that the flux linkage for $n$ turns of the coil is

$$\Phi = n \times B\pi r^2$$

i.e. $n$ times the flux linkage for one turn.

Substituting for $B$ given above, the flux linkage for a long coil is given as

$$\Phi = \frac{\mu_0 \pi n^2 r^2}{l} I$$

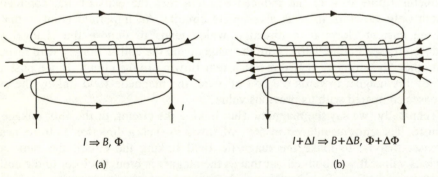

$$I \Rightarrow B, \Phi \qquad\qquad I+\Delta I \Rightarrow B+\Delta B, \Phi+\Delta\Phi$$

(a)                                    (b)

*Figure 10.1* **Changing the current flowing through a coil causes the flux linkage to change**

Note that the flux linkage depends linearly upon the current flowing through the coil. Let us consider what happens if the current through the coil changes. Suppose the current flowing through the coil changes from $I$ to $I + \Delta I$ in the time $\Delta t$. We must have that the magnetic flux linking the circuit changes from $\Phi$ *to* $\Phi + \Delta\Phi$, given as

$$\Phi + \Delta\Phi = \frac{\mu_0 \pi n^2 r^2}{l} (I + \Delta I)$$

See Figure 10.1.

The important point to note is that there has been a change in the flux linkage to the coil when the current changes. The change in the flux linkage, $\Delta\Phi$, is determined from the above in a straightforward manner as

$$\Delta\Phi = \frac{\mu_0 \pi n^2 r^2}{l} \Delta I$$

Let us now divide both sides by $\Delta t$, the time taken for the change. We have

$$\frac{\Delta\Phi}{\Delta t} = \frac{\mu_0 \pi n^2 r^2}{l} \frac{\Delta I}{\Delta t} \tag{10.3}$$

But, according to (9.4), $\Delta\Phi/\Delta t$ is just the magnitude of the induced e.m.f. when the flux linking a conductor changes, i.e. Faraday's law. We therefore conclude that when we change the current through a coil, an e.m.f. is induced in the coil. The magnitude of this induced e.m.f. is given by Equation (10.3), and depends not only on how quickly we try to change the current flowing through the coil, but also on the properties of the coil itself.

We note from Lenz's law the polarity of the induced e.m.f. will be such that it opposes the change in the current, so it follows from the above discussion that a coil has a natural property of *opposition* to *changes* in the flow of current. This property is referred to as its

> *self-inductance*

(conveniently referred to just as its *inductance*). It follows from (10.3) that we can write Lenz's law, (9.6), as follows:

$$\mathscr{E} = -\frac{\Delta\Phi}{\Delta t} = -\frac{\mu_0 \pi n^2 r^2}{l} \frac{\Delta I}{\Delta t} \tag{10.4}$$

The above argument is quite general and we conventionally define the inductance by the equation

$$\mathscr{E} = -L \frac{\Delta I}{\Delta t} \tag{10.5}$$

The minus sign shows that the induced e.m.f. opposes the change in the current.

Comparing Equations (10.4) and (10.5), we note the inductance for a long coil is given by the expression

$$L = \frac{\mu_0 \pi n^2 r^2}{l}$$

## Terminology

■ A coil is often referred to as an inductor. The symbol $L$ not only depicts an inductor but also a property referred to as its inductance. Inductance is a property of the inductor that measures the opposition to changes in the flow of current.

■ For the case of the inductor the induced e.m.f. is referred to as the back e.m.f.

## The S.I unit of inductance

The SI unit of inductance is the henry (H). It is defined with respect to the definition (10.5):

**DEFINITION**

> If a current change of one ampere per second induces an e.m.f. of one volt in an inductor, then its inductance is one henry.

We previously quoted the permeability of free space as $\mu_0 = 4\pi \times 10^{-7}$ in SI units. It can be shown that permeability has the SI unit of henry per metre ($H\,m^{-1}$), so that we have

$$\mu_0 = 4\pi \times 10^{-7}\ H\,m^{-1}$$

Worked Examples 10.1 and 10.2 are straightforward numerical exercises that make use of (10.5).

**WORKED EXAMPLE 10.1**

What is the inductance of a coil in which an e.m.f. of 20 V is induced by a uniform current change of 200 mA in 100 ms?

− Here we have $\mathscr{E} = 20$ V, $\Delta I = 200$ mA $= 0.2$ A, $\Delta t = 100$ ms $= 0.1$ s. Making use of (10.5), we have

$$20 = L\,\frac{\Delta I}{\Delta t} = L\,\frac{0.2}{0.1} \rightarrow L = 10\ H$$

Note that we have dropped the minus sign in (10.5). The inductance is a positive number: We are given that the e.m.f. induced is 20 V, this is taken as the magnitude of the induced e.m.f. The polarity of the e.m.f. is taken into account by Lenz's law.

**━━━ WORKED EXAMPLE 10.2**

Determine the magnitude of the back e.m.f. induced in a coil of inductance 10 mH if the current changes uniformly by 50 mA in 5 μ s.

− We are given $L = 10$ mH $= 10 \times 10^{-3}$ H, $\Delta I = 50$ mA $= 50 \times 10^{-3}$ A, $\Delta t = 5$ μ s $= 5 \times 10^{-6}$ s, and we wish to determine the back e.m.f. Making use of (10.5), again ignoring the minus sign since we wish to determine the magnitude of the back e.m.f., we find

$$\mathscr{E} = 10 \times 10^{-3} \times \frac{50 \times 10^{-3}}{5 \times 10^{-6}} \text{ V} = 100 \text{ V}$$

Worked Example 10.3 further emphasizes that it is the change in current which is important.

**━━━ WORKED EXAMPLE 10.3**

It is arranged that the current through an inductor of inductance 1 H has the form shown in Figure 10.2. Determine the e.m.f. induced in the inductor during the times AB, CD and EF.

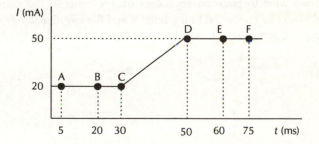

*Figure 10.2*

− During the time AB the current is constant. This means that there is no change in the current flowing through the inductor and, therefore, the e.m.f. induced in the inductor is zero. The same argument can be given for the time EF. For the time CD the current has changed from 20 mA to 50 mA in a time of (50–30) ms = 20 ms. It follows that the magnitude of the induced e.m.f. is given as

$$\mathscr{E} = 1 \text{ Hx} \frac{(50 - 20) \times 10^{-3} \text{ A}}{(50 - 30) \times 10^{-3} \text{ s}} = 1.5 \text{ V}$$

Worked Example 10.4 discusses Lenz's law and again emphasizes that it is the rate at which the current changes which is important, and not the direction of the current flow.

▬▬▬ **WORKED EXAMPLE 10.4**

Determine the polarity of the back e.m.f. in Figures 10.3 and 10.4.

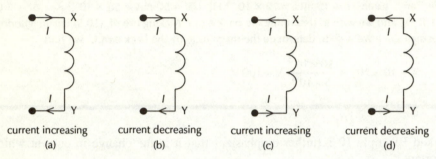

| current increasing | current decreasing | current increasing | current decreasing |
| (a) | (b) | (c) | (d) |

*Figure 10.3*

– In Figure 10.3(a) the current is flowing through the inductor in a direction from X to Y, and we are given that the current is increasing. According to Lenz's law, the back e.m.f. has a polarity such that it opposes the change which caused it. Here, the change is the current increasing, so the polarity of the back e.m.f. is such that the induced current flows from point Y to point X. If we were to place an equivalent battery between the points X and Y, the positive terminal would be connected to the point X and the negative terminal to the point Y, causing the induced current to flow from Y to X; see Figure 10.4.

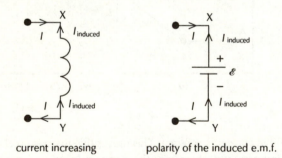

current increasing    polarity of the induced e.m.f.

*Figure 10.4* **When the current through the inductor, *I*, increases the induced current flows in the direction so as to try and reduce the increase**

Similarly, in Figure 10.3(b), we are given that the current is decreasing, so that the induced current flows in a direction such that it tries to keep the current flowing through the inductor at its initial value. The polarity of the induced e.m.f. for Figures 10.3(b), (c) and (d) follow.

## 10.2 Inductance in a circuit

We now consider a circuit consisting of an inductor, $L$, connected in series with a resistor $R$ and a battery of e.m.f. $E$, as shown in Figure 10.5. Also shown in the circuit is an ammeter which we assume is perfect and does not influence the circuit. Note that, since the inductor is made of wire it will have, in general, a non-zero resistance. We can always associate any resistance the inductor has with the resistor $R$, so it follows that we can always assume that the inductor has no resistance.

We wish to determine the current through the circuit after the switch, S, has been closed. After the switch S is closed, the battery is connected to a resistor, $R$, and an inductor, $L$, which we have argued that we may take to have zero resistance, so surely the current is just given by Ohm's law, i.e. $I = E/R$. In fact, we have to be a bit more careful than this because of the presence of the inductor in the circuit.

Before attempting to discuss this circuit, we note that the Ohm's law value of the current, $E/R$, is important and will eventually be interpreted as the final value of the current flowing through the circuit, i.e. the current flowing through the circuit a long time after the switch has been closed.

We first present a qualitative discussion of what we might expect when we close the switch, S. Then we quote the solution and show it has the properties we expect. Later we present a more technical discussion based upon an analysis using Kirchhoff's laws.

### Qualitative discussion

We first consider the times *just before* and *just after* the switch is closed. Just before the switch is closed, the battery is not connected to the circuit, so we must have that the current through the circuit is zero – this is straightforward. What is the current *just after* the switch is closed? A related question is: 'does the current rise *instantaneously* from zero to the Ohm's law value of $E/R$ when S is closed?'. The key to this question is the presence of the inductor in the circuit, which we now explain.

We have previously seen that when a current exists a magnetic field is created. Before S is closed there was no current in the circuit, it follows that there was no

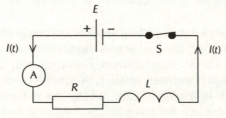

*Figure 10.5* **The effect of inductance in a circuit**

magnetic field linking the inductor. It also follows that, when a current does flow through the circuit, the magnetic field linking the inductor changes from its initial zero value to some non-zero value. Thus, there is a *change* in the magnetic flux linking the inductor.

According to Faraday's law, (9.4), if there is change in flux linking the inductor, $\Delta\Phi$ say, there is an e.m.f. induced in the inductor. If the current were to change instantaneously from its zero value to its final value as we closed the switch, there would be a non-zero change of flux, $\Delta\Phi$, occurring in a zero time ('instantaneously' means that $\Delta t$ is zero). This would mean that the e.m.f. induced in the coil would be infinite, since we have a non-zero quantity, $\Delta\Phi \neq 0$, divided by a zero quantity, $\Delta t = 0$. This is not reasonable, so we are forced to come to the conclusion that when we close S in Figure 10.6, the current does not change instantaneously from its zero value to its final value given by Ohm's law value, $E/R$. This implies that just after the switch has been closed, the current in the circuit remains zero.

It is more constructive to discuss the situation in terms of Equation (10.5), which relates the back e.m.f. generated in the inductor, $\mathscr{E}$, to the inductance, $L$, and the rate of change of the current, $\Delta I/\Delta t$.

Just after the switch has been closed, the back e.m.f. must be equal in magnitude to the e.m.f. of the battery. The p.d. across the resistor is the difference between the e.m.f. of the battery and the back e.m.f. generated across the inductor. Because the polarity of these two e.m.f.'s are opposite (this is just Lenz's law), the net effect is that the p.d. across the resistor is zero. If this is so, then, by Ohm's law, the current through the resistor, and therefore through the circuit, must be zero.

The presence of the back e.m.f. means that, although the current is zero, the rate of change of current is non-zero. In fact, just after S is closed, the rate of change of current is given as $E/L$, since the magnitude of the back e.m.f. $\mathscr{E}$, is equal to the e.m.f. of the battery, $E$, at time $t$ equal to zero (from Equation (10.4)).

If the rate of change of the current is non-zero, it must mean that the current is changing. Therefore, the current begins to rise. If a current flows through the circuit, it also flows through the resistor, so there must be a p.d. across the resistor (recall that current only flows through a resistor if there is a p.d. across the resistor). This must mean that the back e.m.f., $\mathscr{E}$, becomes smaller than the e.m.f. of the battery, $E$. According to (10.5) this means that the rate of change of current, $\Delta I/\Delta t$, becomes smaller than its initial value (recall that if the current is changing at a slower rate, the flux linking the inductor is also changing at a slower rate and the back e.m.f. is smaller).

This process of the back e.m.f. getting smaller as time proceeds continues. As the back e.m.f. becomes smaller, the p.d. across the resistor increases and more current flows through the circuit. As more current flows through the circuit, the rate at which the current changes becomes smaller. A smaller rate of change of current implies that the change in magnetic flux linking the inductor gets smaller, which implies that the induced e.m.f. becomes smaller. After a very long time, we expect the current to reach the steady value given by Ohm's law, i.e. $E/R$. If the current is steady, the magnetic flux linking the inductor does not change, which implies that the back e.m.f. induced across the inductor is zero.

The relationship between the current, the p.d. across the resistor, the rate at which the current changes and the back e.m.f. is illustrated in Figure 10.6.

### The solution

We quote the solution of the current at time $t$ after the switch has been closed as

$$I(t) = \frac{E}{R} \times \left(1 - \exp\left(-\frac{R}{L}\,t\right)\right) \tag{10.7}$$

The exponential function is discussed in the Appendix.

We emphasize that the current depends upon time by writing $I(t)$, which is taken to read the 'current at time $t$'.

We can write (10.7) in a slightly different way by making use of the following

$$\frac{R}{L}\,t = \frac{1}{(L/R)}\,t = \frac{t}{(L/R)} = \frac{t}{\tau}$$

where the symbol $\tau$ (the Greek symbol *tau*) is defined by the equation

$$\tau = \frac{L}{R} \tag{10.8}$$

and is referred to as the time constant, for reasons that will soon become obvious. With the time constant so defined, we can write (10.7) in the form

$$I(t) = \frac{E}{R} \times \left(1 - \exp\left(-\frac{t}{\tau}\right)\right) \tag{10.9}$$

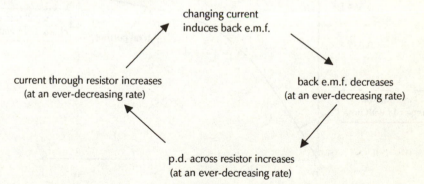

Figure 10.6   **The relationship between changing current and the value of the current**

The variation of the current with time is given in Figure 10.7. The behaviour of the current is in agreement with that expected, i.e. a gradual rise from zero when the switch is closed, to the final current given by Ohm's law for a battery of e.m.f. $E$ connected to a single resistor $R$:

$$I_{max} = \frac{E}{R}$$

Note also that just after the switch has been closed, although the instantaneous value of the current is zero (as expected), the *rate* at which the current is changing is non-zero. This rate of change is largest at time $t$ equal to zero and then gradually gets smaller. As the current reaches its maximum value, so the rate at which the current is changing gets smaller.

As its name suggests, the time constant has the units of time. What does the time constant measure? Note that when $t = \tau$

$$\exp\left(-\frac{t}{\tau}\right) = \exp(-1) = 0.368, \quad \therefore 1 - \exp\left(-\frac{t}{\tau}\right) = 0.632$$

i.e. the time constant measures the amount of time that elapses for the current to *reach* 63.2% (i.e. nearly 2/3) of its final value. Notice that after about five time constants, the current has reached within 1% of the maximum current, given by $I_{max} = E/R$. The time constant, therefore, gives us information on how long we have to wait for the current to reach its final value, and so is appropriately named.

From the definition of the time constant we note that it is proportional to the inductance and inversely proportional to the resistance. Is this dependence on $L$ and

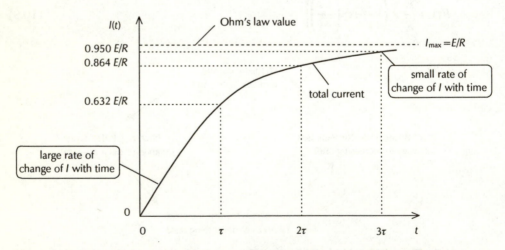

*Figure 10.7*  **The variation of the induced and total current in the circuit with time**

R reasonable? Given that the inductance measures opposition to a change in current, so $\tau$ increasing as $L$ increases is reasonable. This is because we expect to wait longer for the current to reach its final value. Finally, $1/R$ determines the maximum current that flows through the circuit. Since the initial current is zero, $1/R$ also measures the change in the current through the circuit. It follows that we expect $\tau$ to increase with $1/R$, since if a larger current flows through the circuit the change in the current is larger, and we therefore expect to have to wait longer before we reach that final value.

### Energy associated with an inductor

From our previous discussions, it is obvious that we have to do work in changing the current in an inductor from its initial zero value to some particular value that happens to be flowing at time $t$, say $I(t)$. This is because the battery has to do work in driving the current against the back e.m.f. induced in the inductor as the current changes its values. It is argued that this work is stored in the magnetic field of the inductor. Another way to see this is to consider pushing a magnet towards an inductor. Since the current induced in the inductor opposes the change of magnetic flux linking the inductor, work has to be performed to bring the magnet closer to the inductor.

We quote the result that when a current $I$ is flowing in an inductor with inductance $L$, the energy stored in the magnetic field is

$$\tfrac{1}{2}LI^2$$

Worked Example 10.5 brings together the concepts discussed previously.

━━━ **WORKED EXAMPLE 10.5**

Consider the circuit shown in Figure 10.8. Determine the current flowing through the circuit, $I$, the p.d. across the resistor, $V_R$, the power dissipated by the resistor, $P$, the p.d. across the inductor, $V_L$, and the energy stored by the inductor, $E_L$, at the following times: $t = 0$, 0.5, 1.0, 2.5, 5.0 and 10 ms, where the switch S has been closed at time $t = 0$. How long does the current take to reach a value $I = 2$ mA?

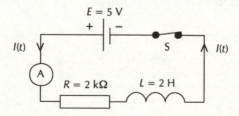

*Figure 10.8*

– The current flowing through the circuit at any time $t$ after the switch has been closed is given by (10.9) (or, equivalently, (10.7)). The time constant for the circuit is given as

$$\tau = \frac{L}{R} = \frac{2H}{2k\Omega} = 1 \times 10^{-3} \, s = 1 \, ms$$

This time sets the scale for the time behaviour of the circuit, and we know that after one time constant the current flowing through the circuit has reached about 63% of its maximum value. Once we have the current, the p.d. across the resistor is given by Ohm's law. With both $I$ and $V_R$ given, the power $P$ dissipated by the resistor is given by their product. The p.d. across the inductor is given by a simple application of Kirchhoff's law once the p.d. across the resistor is known. If we call $V_L$ the p.d. across the inductor, then according to Kirchhoff we have at each instant in time

$$E = V_R + V_L$$

This point will be discussed in some detail in Section 10.3. Finally, once the current has been determined, the energy stored by the inductor is known. Our results are presented in Table 10.1.

Table 10.1

| $t$ (ms) | $I$ (mA) | $V_R$ (V) | $P$ (mW) | $V_L$ (V) | $E_L$ (μJ) |
|---|---|---|---|---|---|
| 0.0 | 0.0 | 0.0 | 0.0 | 5.000 | 0.000 |
| 0.5 | 0.984 | 1.967 | 1.935 | 3.033 | 0.968 |
| 1.0 | 1.580 | 3.161 | 4.995 | 1.839 | 2.496 |
| 2.5 | 2.295 | 4.590 | 10.532 | 0.410 | 5.267 |
| 5.0 | 2.483 | 4.966 | 12.332 | 0.034 | 6.165 |
| 10.0 | 2.500 | 5.000 | 12.499 | 0.000 | 6.250 |

Note that, once given the time it is straightforward to determine the current. In the next part of the question we are asked to determine the time for which a particular current is flowing. We have to solve (10.9) with the following information:

$$2.0 = 2.5 \times \left(1 - \exp\left(-\frac{t}{\tau}\right)\right), \quad \text{with } \tau = 1 \text{ ms}$$

i.e. we want the value of $t$ which gives $I(t) = 2$ mA. If we multiply the above equation out we can eventually rewrite it as follows:

$$2.5 \exp\left(-\frac{t}{\tau}\right) = 2.5 - 2.0 = 0.5$$

from which we have

$$\exp\left(-\frac{t}{\tau}\right) = \frac{0.5}{2.5} = 0.2$$

From this equation we find

$$\frac{t}{\tau} = -\ln(0.2) = 1.609$$

Some details of the logarithm function are given in the Appendix. Since $\tau$ is given as 1 ms, we have the required time given as

$$t = 1.609 \text{ ms}$$

---

## ○ 10.3 Analysis based on Kirchhoff's laws

We now provide a technical discussion based upon Kirchhoff's laws. Consider Figure 10.9, which shows a battery of e.m.f. $E$ connected in series with a resistance $R$ and an inductance $L$. In the figure the arrow notation introduced earlier is also indicated. The switch, S, has been closed for a time $t$. At this time take the current through the circuit to be $I$, and the rate at which this current is changing to be $dI/dt$. The p.d. across the resistor is $V_R$, while the back e.m.f. developed across the inductor is $\mathscr{E}$. We have

$$V_R = IR, \qquad \mathscr{E} = -L\frac{dI}{dt}$$

Note that we have to make full use of calculus notation and have replaced

$$\frac{\Delta I}{\Delta t} \rightarrow \frac{dI}{dt}$$

Applying Kirchhoff's voltage law to the circuit depicted in Figure 10.9, we have

$$E - V_R + \mathscr{E} = 0$$

Substituting for $V_R$ and $\mathscr{E}$ and rearranging, we find we can write the above as a first-order linear differential equation:

$$\frac{dI}{dt} + \frac{R}{L}I = \frac{E}{L}$$

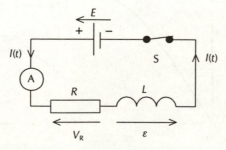

*Figure 10.9* **Inductance in a circuit and Kirchhoff's laws**

The solution of this differential equation is

$$I(t) = \frac{E}{R} \times \left(1 - \exp\left(-\frac{R}{L}t\right)\right) \tag{10.10}$$

where we have made use of the boundary condition that there is no current through the circuit at $t = 0$, when the switch S is closed.

We can interpret the solution (10.10) in terms of the Principle of Superposition (see Section 7.2) as follows: The current through the circuit is made up of the steady state current due to the battery in series with the resistor $R$ alone together with the induced current due to the back e.m.f. If we write the induced current, $I_{\text{induced}}(t)$, through the circuit at time $t$ after the switch has been closed as

$$I_{\text{induced}}(t) = \frac{E}{R} \exp\left(-\frac{R}{L}t\right)$$

it follows that the total current, $I$, through the circuit at time $t$ is given as

$$I(t) = \frac{E}{R} - I_{\text{induced}}(t) = \frac{E}{R} \times \left(1 - \exp\left(-\frac{R}{L}t\right)\right)$$

which is the same as Equations (10.8) and (10.10). That the currents subtract from each other is a consequence of Lenz's law: the change is the circuit current changing from its initial value of zero to its final value of $E/R$. It follows that the induced current flows in the direction opposite to the current supplied by the battery, trying to keep the total circuit current at its initial value. The variations of the induced and total currents with time are given in Figure 10.10.

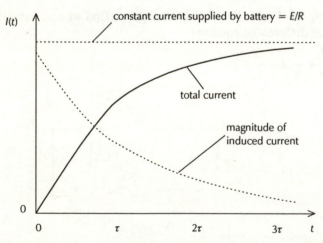

*Figure 10.10*  **The variations of the induced and total currents in the circuit with time**

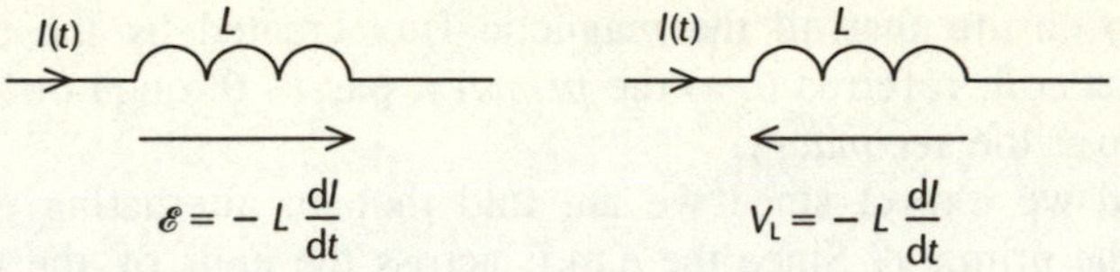

$$I(t) \xrightarrow{\quad} \ \overset{L}{\text{⌒⌒⌒}} \ \underline{\qquad} \qquad\qquad I(t) \xrightarrow{\quad} \ \overset{L}{\text{⌒⌒⌒}} \ \underline{\qquad}$$

$$\xrightarrow{\hspace{2cm}} \qquad\qquad\qquad \xleftarrow{\hspace{2cm}}$$

$$\mathscr{E} = -L\,\frac{\mathrm{d}I}{\mathrm{d}t} \qquad\qquad\qquad V_L = -L\,\frac{\mathrm{d}I}{\mathrm{d}t}$$

*Figure 10.11* **The relationship between back e.m.f. and p.d. across an inductor**

We can derive the following result from (10.8):

$$\frac{\mathrm{d}I}{\mathrm{d}t} = \frac{E}{L}\exp\!\left(-\frac{t}{\tau}\right)$$

from which we can derive the back e.m.f. generated across the inductor as

$$\mathscr{E} = -L\,\frac{\mathrm{d}I}{\mathrm{d}t} = -E\exp\!\left(-\frac{t}{\tau}\right)$$

Note that as $t$ gets very large, $-L\,\mathrm{d}I/\mathrm{d}t$ tends to zero, i.e. the current reaches a steady value, and therefore the back e.m.f. tends to zero.

We note, in passing, that instead of talking of the back e.m.f., $\mathscr{E}$, developed across the inductor, it is sometimes more convenient to speak of the p.d., $V_L$, across the inductor. The relationship between the two quantities are given in (10.11) and in Figure 10.11.

$$\mathscr{E} = -L\,\frac{\mathrm{d}I}{\mathrm{d}t}, \qquad V_L = L\,\frac{\mathrm{d}I}{\mathrm{d}t} \tag{10.11}$$

## 10.4 The principle of the transformer

Consider Figure 10.12, which shows two sets of insulated coils surrounding an iron core. Through one of the sets of coils we pass an *alternating* e.m.f. The role of the

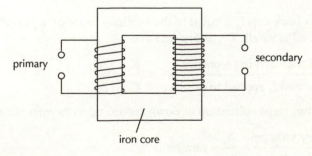

*Figure 10.12* **The transformer**

iron core is to ensure that all the magnetic flux created by the current flowing through the first coil, referred to as the *primary*, passes through or links the second coil, referred to as the *secondary*.

What should we expect since we are told that an alternating e.m.f. has been connected to the primary? Since the e.m.f. across the ends of the primary is time dependent, this means that the current through the coil is time dependent. It then follows that the magnetic field created by the coil is changing with time. Thus the magnetic flux linking the secondary coil is changing. But if the magnetic flux linking this coil is changing then we know that an e.m.f. is induced in the secondary.

We will assume that the primary coil is long. In this case the magnetic field created by the current through the primary coil is uniform inside and zero outside (recall the work in Section 8.2). Suppose the current flowing through the primary circuit is such that the magnetic field strength inside the coil is $B$. We will assume that the primary coil has a cross-sectional area $A$. It follows that the magnetic flux $\Phi$ linking the coil is given as

$$\Phi_p = N_p BA$$

where $N_p$ is the number of primary turns.

Because of the iron core, all this changing magnetic field links the secondary coil which, therefore, induces an alternating e.m.f. $\mathscr{E}_s$. Suppose the number of turns of the secondary is $N_s$, then the flux linking the secondary is

$$\Phi_s = N_s BA$$

The magnitude of the e.m.f. induced in the secondary is given from Faraday's law (see Equation (9.4)):

$$\mathscr{E}_s = \frac{d\Phi_s}{dt} = N_s A \frac{dB}{dt}$$

We know that this changing flux generates a back·e.m.f. in the primary with magnitude given by

$$\mathscr{E}_p = \frac{d\Phi_p}{dt} = N_p A \frac{dB}{dt}$$

We note that this back e.m.f. is equal to the applied source of e.m.f. if the resistance of the primary coil is small. Therefore we have

$$\frac{\text{e.m.f. induced in secondary}}{\text{source of e.m.f. applied to primary}} = \frac{\mathscr{E}_s}{\mathscr{E}_p} = \frac{N_s}{N_p}$$

Thus a transformer steps voltage up or down according to its *turn ratio*:

$$\frac{\text{secondary voltage}}{\text{primary voltage}} = \frac{N_s}{N_p} = \text{turn ratio}$$

The transformer is referred to as a step-up transformer if the number of secondary turns is larger than the number of primary turns, i.e. $N_s > N_p$, since then the secondary voltage is larger than the primary voltage. Otherwise, the transformer is referred to as a step-down transformer.

Note that transformers only work when the voltage applied across the primary changes with time.

### Uses of transformers

One of the main uses of transformers is the transmission of electrical power over large distances with very little loss of energy. This is a major advantage of AC over DC power transmission, and means that it is feasible to produce electricity in one location and to convey it to another location via high-voltage power lines. This system of power lines is referred to as the *National Grid*, or simply the *Grid*.

Electricity is generated in power stations at 11–33 kV and is then stepped up to 400 kV by transformers. It is fed into the grid at this voltage and subsequently stepped down in successive stages at substations in the neighbourhood of towns where the energy is to be used.

Worked Example 10.6 shows that a much smaller power loss occurs in the power lines when energy is transmitted at a high voltage.

---

■■■■ **WORKED EXAMPLE 10.6**

---

Determine the power dissipated as heat in a cable when 10 kW is transmitted through a cable of resistance 0.5 Ω at 200 V and at 200 kV.

– At 200 V the current through the cable is

$$I = \frac{P}{V} = \frac{10\,000}{200} \text{ A} = 50 \text{ A}$$

and the power loss in the cable is

$$I^2R = 50^2 \times 0.5 \text{ W} = 1.25 \text{ kW}$$

If the power is transmitted at 200 kV the current through the cable is

$$I = \frac{P}{V} = \frac{10\,000}{200\,000} \text{ A} = 0.05 \text{ A}$$

giving the power loss in the cable as

$$I^2R = 0.05^2 \times 0.5 \text{ W} = 0.001\,25 \text{ W}$$

At 200 V, therefore, more than 10% of the energy is wasted merely in warming the cable, whereas at 200 kV the energy losses are negligible.

---

### 10.5 Summary

We discussed the natural property of a coil (technically referred to as an inductor) called its 'inductance' or more accurately its 'self-inductance', of opposing *changes* to the flow of current in great detail.

We found that it is the *rate* at which current changes, as distinct from the current itself, that determines the size of the back e.m.f. induced in the inductor. In particular, the size of the back e.m.f. is directly proportional to the rate of change of current. The property which links these two quantities is the inductance of the coil.

When an inductor is connected in series to a resistor and a battery we found that the inductance has a dramatic effect on the current through the circuit. We found that the current through the circuit depends upon the time after the switch is closed. This is directly related to the property of the inductance.

Given that we do work to change the current through an inductor from its initial zero value to some particular value that happens to be flowing at time $t$, say $I(t)$, it is not surprising that inductors can be used to store energy. This energy is stored in the magnetic field of the inductor.

As a practical use of inductors we discussed transformers. We showed that they can be used to either step-up or step-down voltages. This has enormous practical use in transmitting electrical energy from one place to another.

### ▬ Problems

In the following take the permeability of free space $\mu_0$ as $4\pi \times 10^{-7}$ H m$^{-1}$.

**10.1**   Determine the inductance of an inductor if an e.m.f. of 10 V is induced by a uniform current change of 100 mA in 5 ms.

**10.2**   A current changes uniformly in an inductor by 5 mA in a time of 10 ms. If the back e.m.f. is 50 V what is the inductance? If the current changes by 20 mA in 5 ms, what back e.m.f. is induced?

**10.3**   Determine the inductance of a long coil of radius $r = 2$ mm, length 100 mm and total number of turns 100.

o **10.4**   Confirm that the time constant, $L/R$, has the units of time.
   *Hint:* Make use of the results of Chapter 1.

**10.5**   Consider the circuit shown in Figure P10.1.

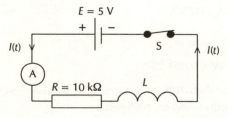

*Figure P10.1*

(a) Given that a back e.m.f. of 10 V is induced in the inductor if the current changes uniformly by 100 mA in a time 10 ms, determine the inductance, $L$. The switch S is closed at time $t = 0$.

(b) Determine the maximum current flowing.

(c) Determine the time constant, $\tau$.

(d) How long does the current take to reach a value $I = I_{max}/2$?

(e) What is the energy stored in the inductor at this time?

(f) What is the instantaneous power dissipated by the resistor at this time?

(g) Complete Table P10.1. In the table, $I$ is the current flowing in the circuit, $V_R$ and $V_L$ are the potential differences across the resistor and the inductor, respectively, $P$ is the power dissipated by the resistor, and $dI/dt$ is the rate of change of current, all at time $t$. The question relating to the rate of change of current may be taken as advanced and ignored if desired.

Table P10.1

| t (ms) | I (mA) | $V_R$ (V) | P (mW) | $V_L$ (V) | dI/dt (mA/s) |
|---|---|---|---|---|---|
| 0 | | | | | |
| 0.05 | | | | | |
| 0.15 | | | | | |
| 0.25 | | | | | |
| 0.50 | | | | | |
| 1.00 | | | | | |

*Hint:* If the current flowing through the circuit is known, the p.d. across the resistor is given as $IR$. The p.d. across the inductor, $V_L$, is given as $L\, dI/dt$. Alternatively, from Kirchhoff, if $V_R$ is known, then $V_L$ is also known. Once $V_L$ is known, $dI/dt$ can be determined.

(h) Plot your results for $I$ and $dI/dt$ on a graph and comment on your results.

**10.6** A transformer has a primary coil consisting of 100 turns and a secondary of 250 turns. Does the transformer step-up or step-down voltages? If a 100 V DC source is connected across the primary coil what is the voltage across the secondary? If a 250 V AC source is connected across the primary coil what is the voltage across the secondary? Assume the transformer is perfect.

**10.7** An AC voltage source is connected across the primary which consists of 100 turns. If the secondary coil consists of 300 turns and the output voltage across it is 200 V, what is the voltage across the primary?

# CHAPTER ELEVEN

# *Capacitance*

In this chapter we discuss the concept of capacitance. We will find that capacitors are devices used for storing electric charge and electrical energy. Some discussion will be presented for the most common form of capacitor, namely a pair of conducting plates separated by an insulating material.

We will discuss how capacitors can be arranged in series and parallel and we will derive formulae which determine the effective capacitance of such arrangements.

If the capacitor is originally uncharged this means that there must, at some stage, be a flow of charge, i.e. an electric current must exist. How this flow of charge is achieved and its time dependence will be a major topic of this chapter.

Finally, it will be noted that as a capacitor is being charged, it is necessary to do work to move like charges closer to each other. It follows that when charged, the capacitor stores energy. This energy is said to reside in the electric field between the plates.

## 11.1 **Introduction to capacitance**

A capacitor is a device for storing electric charge. In its simplest and most widely used form it consists of a pair of parallel conducting plates placed a small distance apart. Usually the plates are separated by an insulating material referred to as a dielectric. This arrangement is referred to as the parallel plate capacitor.

Let us consider the action of connecting a battery to opposite plates of an initially uncharged capacitor, as shown in Figure 11.1. Recall that a battery can be thought to act as an electron pump, pushing out negative electrons at the negative terminal and accepting the electrons at the positive terminal. If a battery is connected to the capacitor as shown in the figure, electrons are pushed out from the negative terminal onto the lower plate. These electrons are unable to flow across the dielectric since, by definition, it is made of an insulating material, and so they accumulate on the plate, which therefore acquires a negative charge. Although these electrons cannot move across the plate, their (electrical) *influence* can be felt by electrons on the

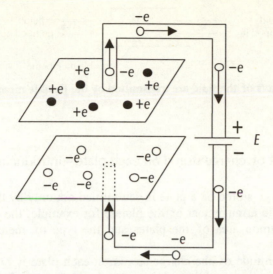

*Figure 11.1*    **Connecting a source of e.m.f. to a parallel plate capacitor**

upper plate. Electrons on the upper plate are repelled by the electrons on the lower plate. Since electrons move away from the upper plate it is left with a positive charge (since there now is a deficiency of negative charge on this plate). Since charge is in motion we can appreciate that an electric current momentarily flows.

The electron flow (and therefore the current) will not continue indefinitely. This is because as the charge acquired by each plate increases, so too does the p.d. between them. The charging process finishes when the p.d. across the plates is the same as the e.m.f. of the battery.

Note that the lower plate acquires a negative amount of charge and the upper plate acquires an equal but opposite positive amount of charge, so that if the amount of charge acquired by the lower plate is $-Q$ the upper plate acquires an amount of charge $+Q$. When we refer to the charge on a capacitor we usually just quote the magnitude of the charge on each plate.

Again, we emphasize that there is no current flow *between* the plates – there cannot be since they are separated by an insulating material – but the effects of the charge are passed across the plates. A useful analogy is often made to a pipe with a flexible membrane between the two ends of the pipe; see Figure 11.2. Here, fluid is pumped in through one end of the pipe. Because of the membrane there can be no transfer of fluid. The effects of force transmitted by the fluid at one end of the pipe are transmitted to the other end of the pipe via the membrane. This point will discussed in more detail in Section 11.4.

An important point to note is that the charges stored on the plates remain even when the battery is removed, until a conductive path is provided between the plates. In this sense we can see that the capacitor is a device that can store electric charge.

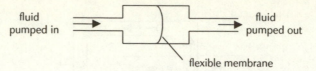

*Figure 11.2* **The *effects* of the fluid are transmitted by the flexible membrane but there is no transfer of fluid**

## The amount of charge stored on each plate – introducing capacitance

The amount of charge stored on a plate is determined not only by the p.d. across the plates but also by the arrangement of the plates, for example, the distance between the plates, the common area of the plates and the type of material between the plates.

Suppose the magnitude of the charge stored on each plate is $Q$, i.e. $+Q$ on one plate and $-Q$ on the other, and the p.d. between the plates is $V$. Experimentally it is found that the amount of charge stored is proportional to the p.d. between the plates:

$$Q \propto V \tag{11.1}$$

This is perhaps reasonable, since, referring to Figure 11.1, we regard e.m.f. as a measure of the ability to move charge completely around the circuit.

The proportionality (11.1) is turned into an equality by introducing a constant, $C$, called the capacitance.

$$Q = C \times V \tag{11.2}$$

The term 'constant' really means 'constant for given arrangement of plates' and, in general, 'constant for given arrangements of conductors'. This will become obvious later.

Capacitance is said to be a measure of the ability of a capacitor to store charge. This is reasonable, since, according to Equation (11.2), for a given p.d. across the plates the amount of charge stored by the capacitor increases as the capacitance, $C$, increases. Evidently, the capacitance depends upon the distance between the plates, the common area of the plates and the type of material between the plates. This will be discussed in Section 11.2.

Often we see Equation (11.2) written in the equivalent form

$$C = \frac{Q}{V} \tag{11.3}$$

From this formula we see that the capacitance may be *defined* as the ratio of the amount of charge stored by a body to the p.d. developed across the body, making a slight generalization from the parallel plate arrangement.

### The SI unit of capacitance

The SI unit of capacitance is the farad (F). It is defined with respect to Equation (11.3) as follows:

■■■■ **DEFINITION**

> If a body, upon acquiring a charge of one coulomb, causes a p.d. of one volt to be developed across the body, then the capacitance is one farad.

The farad represents an enormous capacitance; in practice we meet capacitances of the order

$$\mu F, \quad nF, \quad pF$$

We note that the SI unit of permittivity may be expressed in terms of the unit of capacitance, i.e. the farad, as the farad per metre, i.e. $F\,m^{-1}$ so that the permittivity of free space, $\varepsilon_0$, is

$$\varepsilon_0 = 8.854 \times 10^{-12}\,F\,m^{-1}$$

Worked Examples 11.1 and 11.2 are simple manipulations of Equation (11.3).

■■■■ **WORKED EXAMPLE 11.1**

A battery of e.m.f. 10 V is connected across a 2 μF capacitor. When fully charged, what is the magnitude of the charge on each plate of the capacitor?
– Use $Q = C \times V$ with $C = 2$ μF and $V = 10$ V

$$Q = (2 \times 10^{-6}\,F) \times (10\,V) = 20 \times 10^{-6}\,C = 20\,\mu C$$

Worked Example 11.2 shows that the only difficulty is to determine the charge stored on each plate.

■■■■ **WORKED EXAMPLE 11.2**

A constant current of 5 mA charges a 10 μF capacitor for 1 s. What is the p.d. developed across the capacitor?
– First determine the stored charge. Since we are told that the charging current is *constant*, the amount of charge stored is

$$Q = I \times t = (5 \times 10^{-3}\,A) \times (1\,s) = 5 \times 10^{-3}\,C$$

We now make use of the formula $C = Q/V$ to find

$$V = \frac{Q}{C} = \frac{5 \times 10^{-3}\,C}{10 \times 10^{-6}\,F} = 500\,V$$

## 11.2   **The parallel plate capacitor**

We have mentioned the parallel plate capacitor as the simplest and most widely used form of capacitor. For completeness we discuss the formula for the capacitance of this arrangement.

Figure 11.3 shows the arrangement for the parallel plate capacitor separated by an insulating material. The capacitance, $C$, is given by the equation

$$C = \varepsilon_r \varepsilon_0 \, \frac{A}{d} \tag{11.4}$$

where $A$ is the common cross-sectional area of the plates and $d$ is their separation. The constant $\varepsilon_0$ has been met previously in the discussion of Coulomb's law and the forces between charges (see Section 2.2). This is the permittivity of free space, and we argued that it was a measure of the 'ability' or 'ease' with which electric forces are transmitted between charges in free space. It should not be too surprising that it turns up in capacitance, since capacitors are devices for storing electric charge, this ability being determined by the force between charges. The factor $\varepsilon_r$ is referred to as the relative permittivity or the dielectric constant of the material separating the plates. We note that the relative permittivity is a pure number, and for a vacuum is unity. The value of the relative permittivity for all other materials is larger than this. Hence, placing any (insulating) material between the plates enables us to increase the capacitance compared with the free space value by the factor equal to the relative permittivity, $\varepsilon_r$.

### Some practical aspects of capacitors

We first discuss capacitors which have capacitances that cannot be changed and are referred to as 'fixed capacitors'. We then refer briefly to capacitors for which it is possible to change the capacitance. Such capacitors are termed 'variable capacitors'.

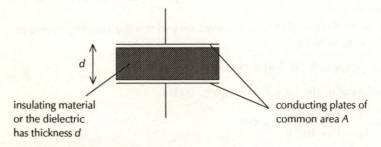

*Figure 11.3*   **The parallel plate capacitor**

### Fixed capacitors

Capacitors are usually classified according to the type of dielectric material. The most common types of dielectric materials are mica, ceramic, plastic film and electrolytic.

*Mica capacitors.* Two types of mica capacitor are the stacked-foil and silver-mica. Such a capacitor consists of alternate layers of metal foil and thin sheets of mica. The metal foil forms the plate, with alternate foil sheets connected together to increase the plate area. As we increase the number of layers, the plate area is increased and therefore the capacitance is increased. The mica and foil stack is then compressed to reduce the physical size and encapsulated in an insulating material. The silver-mica capacitor is formed in a similar manner by stacking mica sheets with silver electrode material screened on them. Mica capacitors are available with capacitance values ranging from 1 pF to 0.1 µF. Mica has a relative permittivity of about 5.

*Ceramic capacitors.* Ceramic dielectrics provide very high relative permittivities ($\varepsilon_r$ around 1200 is typical). For this reason, a comparatively large capacitance can be achieved in a small physical size. Ceramic capacitors typically are available in capacitance values ranging from 1 pF to 2.2 µF.

*Plastic film capacitors.* There are several types of plastic film capacitors that differ according to the particular plastic used as the dielectric. Some of the more common plastics used are polycarbonate, parylene, polyester, polystyrene, polypropylene and mylar. A thin strip of plastic film dielectric is sandwiched between two thin metal strips that act as plates. One of the leads is connected to the inner plate and one to the outer plate. The strips are then rolled in a spiral configuration and encapsulated in a moulded case. Using this method, a large plate area can be packaged into a relatively small physical size. Capacitances up to 100 µF can be achieved in this manner.

*Electrolytic capacitors.* Electrolytic capacitors are available in two types: aluminium and tantalum. The capacitor consists of two strips of either aluminium or tantalum foil separated by a paper or gauze strip saturated with an electrolyte. During manufacture it is arranged for an electrochemical reaction to take place that causes an oxide layer (either aluminium oxide or tantalum oxide) to form on the inner surface of one of the capacitor plates. This oxide acts as the dielectric. Owing to the extreme thinness of the film high capacitance values around 200 000 µF can be achieved.

### Variable capacitors

As their name suggests, it is possible to change the capacitance of these capacitors. Their value is varied by altering the overlap area between a fixed set of metal plates

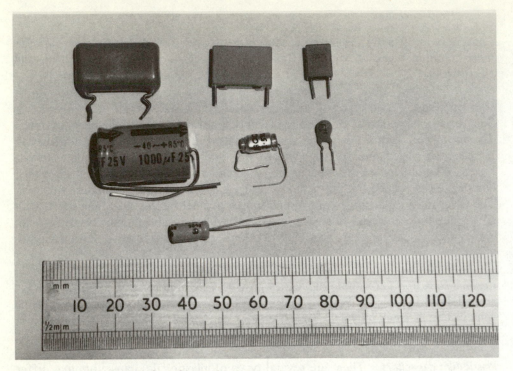

*Figure 11.4* **Some examples of capacitors**

and a moving set, separated by dielectric of air. Small variable capacitors called trimmers or presets are used to make fine, infrequent changes to the capacitance of a circuit. This may be achieved, for example, by compressing a mica capacitor with a screw.

Some real capacitors are shown in Figure 11.4.

## 11.3 Capacitors in series and in parallel

Just as simple resistor networks can be analyzed in terms of their component series and parallel arrangements, so can capacitor networks. It is therefore necessary to determine the rule for the effective capacitance for capacitors in series and in parallel. Again, we start from the simple observation that given a single capacitor of capacitance $C$ it is straightforward to determine the charge on each plate when connected to a source of e.m.f. In this case, the p.d. across the capacitor is equal to the e.m.f. of the source, when fully charged.

We now discuss series and parallel connected capacitors in turn.

## Capacitors in series

Consider the capacitors $C_1$ and $C_2$ connected in series as depicted in Figure 11.5. Suppose a battery is placed across the ends of the capacitor arrangement. The battery acts as an electron pump, electrons are pushed out from the negative terminal onto the left-hand plate of capacitor $C_1$. According to previous arguments, the electrons cannot cross over to the right-hand plate since there is an insulating material between the plates, and so the left-hand plate acquires an amount of negative charge of, say, $-Q$. The electrons on the left-hand plate of capacitor $C_1$ repel an equal number of electrons from the right-hand plate, leaving the charge on the right-hand plate of the capacitor as $+Q$. What happens to the electrons that were pushed away from the right-hand plate? Quite simply, they accumulate on the left-hand plate of capacitor $C_2$ which, therefore, acquires a charge $-Q$. In a similar fashion these electrons repel an equal number of electrons from the right-hand plate of the capacitor which are returned to the battery at the positive terminal, the right-hand plate acquiring a charge $+Q$.

Suppose the p.d. developed across $C_1$ is $V_1$ and across $C_2$ is $V_2$, respectively. Then, from the definition of capacitance (11.3), we have

$$V_1 = \frac{Q}{C_1} \quad \text{and} \quad V_2 = \frac{Q}{C_2}$$

since the charge on both capacitors is $Q$.

Because the capacitors are connected in series, the p.d. across the whole arrangement is the sum of the p.d.'s across each capacitor, i.e.

$$V = V_1 + V_2$$

Substituting for $V_1$ and $V_2$ from above we can write

$$V = \frac{Q}{C_1} + \frac{Q}{C_2}$$

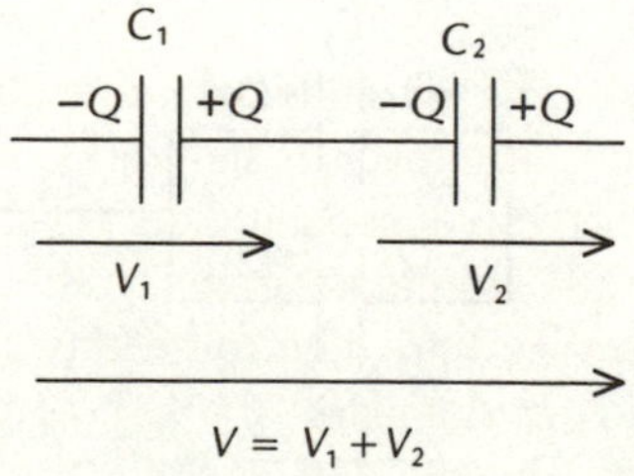

*Figure 11.5* **Capacitors connected in series**

If we define $Q/V$ as the effective capacitance, $C$, of the capacitors $C_1$ and $C_2$ in series, we have

$$\frac{1}{C} = \frac{1}{C_1} + \frac{1}{C_2} \qquad (11.5)$$

For two capacitors only we can write (11.5) in the more useful form

$$C = \frac{C_1 \times C_2}{C_1 + C_2} = \frac{\text{product of capacitances}}{\text{sum of capacitances}} \qquad (11.6)$$

## Capacitors in parallel

Consider the capacitors $C_1$ and $C_2$ connected in parallel, as depicted in Figure 11.6. Again, suppose that we connect a battery across the capacitor arrangement: Electrons are pushed out from the negative terminal of the battery and accumulate on the left-hand plates of the capacitors $C_1$ and $C_2$. Note that we do not necessarily expect that the same number of electrons will be transferred to the plate of each capacitor $C_1$ as to the plates of capacitor $C_2$.

In a manner as described previously, the electrons on the left-hand plate of $C_1$ repel an equal number of electrons on the right-hand plate of $C_1$. Similarly, the electrons on the left-hand plate of $C_2$ repel an equal number of electrons from the right-hand plate of $C_2$. Suppose that the charge on the left-hand plate of capacitor $C_1$ is $-Q_1$ and that on the left-hand plate of capacitor $C_2$ is $-Q_2$. Since we have argued that we do not expect each left-hand plate to have acquired the same number of electrons from the battery, it follows that we do not expect the capacitors to acquire the same charge.

Since we have assumed that the charge acquired by capacitor $C_1$ is $Q_1$ and by $C_2$ is $Q_2$, respectively, then from the definition of capacitance (see Equation (11.3)), we have

$$Q_1 = C_1 \times V \quad \text{and} \quad Q_2 = C_2 \times V$$

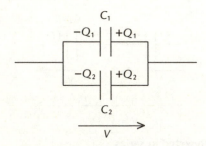

*Figure 11.6*  **Capacitors connected in parallel**

since the capacitors are in parallel and so the p.d. developed across both is the same, call it $V$. In this case the total charge acquired by the system is the sum of the charges on each capacitor. If we were to replace the capacitors $C_1$ and $C_2$ by a single effective capacitor, say $C$, then the total charge on this capacitor is

$$Q = Q_1 + Q_2$$

Substituting for $Q_1$ and $Q_2$ from above yields

$$Q = C_1 \times V + C_2 \times V$$

Dividing by $V$, the p.d. across the effective capacitor, we obtain the formula for the combined or effective capacitance of the capacitors $C_1$ and $C_2$ connected in parallel as

$$C = C_1 + C_2 \qquad\qquad (11.7)$$

■■■ **COMMENTS**

- Notice that for capacitors placed in series, the charge on each of the capacitors is the same, while the p.d.s across them are different. Conversely, capacitors placed in parallel have the same p.d. across them but have different charges.
- Note that in Figures 11.5 and 11.6 we have made use of the voltage arrow notation.
- The formula for capacitors in series is similar to that for resistors in parallel, while that for capacitors in parallel is similar to that for resistors in series.

■■■ **WORKED EXAMPLE 11.3**

Determine the p.d. across each capacitor in the arrangement shown in Figure 11.7(a).

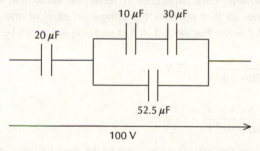

*Figure 11.7(a)*

– The figure shows four capacitors with a p.d. of 100 V across the arrangement. Our first step is to simplify the capacitor network step by step, converting to a single capacitor with a p.d. across its plates of 100 V. We will then be able to determine the charge on each plate of this effective capacitor. If we replace the 10 μF and 30 μF series capacitors by an equivalent 7.5 μF capacitor, we have the network shown in Figure 11.7(b).

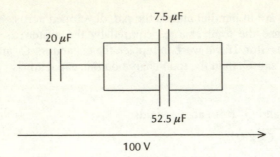

*Figure 11.7(b)*

The 7.5 µF and the 52.5 µF capacitors are in parallel and may be replaced by a 60 µF capacitor, as shown in Figure 11.7(c).

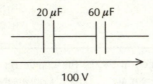

*Figure 11.7(c)*

Finally, we are left with a 20 µF capacitor in series with a 60 µF capacitor. The effective capacitance of this arrangement is 15 µF. From what we mean by the term 'effective capacitance' we know that the p.d. across the 15 µF capacitor is 100 V. It follows that the charge on this effective capacitor is

$$Q = C \times V = (15 \times 10^{-6} \text{ F}) \times (100 \text{ V}) = 1.5 \text{ mC}$$

We now work backwards and determine how this charge is distributed through the original network. From the derivation for capacitors in series we know that the charges on series capacitors are the same, so that we have the charge on each of the 20 µF and the 60 µF capacitors is 1.5 mC. It follows that the p.d. across each capacitor can be determined. We find

$$V_{(20\,\mu F)} = \frac{Q}{C} = \frac{1.5 \times 10^{-3} \text{ C}}{20 \times 10^{-6} \text{ F}} = 75 \text{ V}$$

$$V_{(60\,\mu F)} = \frac{Q}{C} = \frac{1.5 \times 10^{-3} \text{ C}}{60 \times 10^{-6} \text{ F}} = 25 \text{ V}$$

The 60 µF capacitor was the effective capacitance of the 7.5 µF and the 52.5 µF parallel capacitors. The p.d. across the 60 µF capacitor has been determined as 25 V, so that this is the p.d. across each of the 7.5 µF and the 52.5 µF capacitors. Since we know the p.d. across each capacitor we can determine the charge on each capacitor. We find

$$Q_{(7.5\,\mu F)} = C \times V = 0.1875 \text{ mC}$$

$$Q_{(52.5\,\mu F)} = C \times V = 1.3125 \text{ mC}$$

Finally, the 7.5 µF capacitor was the effective capacitance of a 10 µF and 30 µF. We know what the charge on each capacitor is, so we can determine the p.d. across each capacitor. We find

$$V_{(10\,\mu F)} = 18.75 \text{ V} \quad \text{and} \quad V_{(30\,\mu F)} = 6.25 \text{ V}$$

## 11.4  Charging a capacitor

Consider the circuit depicted in Figure 11.8, which shows a battery of e.m.f. $E$ connected in series to a switch, S, a capacitor $C$ and a resistor $R$. Also shown in the circuit is an ammeter which we assume is perfect and does not influence the circuit. The switch is closed at time $t = 0$ and we wish to determine the current that flows through the circuit for times $t$ after this. It is assumed that the capacitor is initially uncharged.

We first discuss in some detail how it is that a current can flow in the above circuit when the switch is closed, since there is no complete path for the current to flow through, recall the parallel plate capacitor consists of two parallel plates separated by an *insulating* material.

After this we ask the question: 'How does the current through the circuit and the p.d. across the capacitor change with time as the switch is closed?' We present a qualitative discussion of what we might expect when we close the switch, S. This is followed by a more technical discussion based on an analysis using Kirchhoff's laws.

### How can current flow through the circuit?

The answer to this is that current flows any time there is a change in the p.d. across the capacitor. To see this, use the formula relating the charge stored on the capacitor plates to the p.d. developed across them:

$$Q = C \times V$$

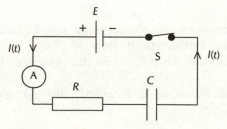

*Figure 11.8*  **Charging a capacitor**

Suppose the p.d. across the capacitor changes from $V$ to $V + \Delta V$ in a time $\Delta t$. If the p.d. has changed, the new charge stored on the capacitor plates is simply

$$Q + \Delta Q = C \times (V + \Delta V)$$

Thus, the charge stored by the capacitor must have changed by an amount $\Delta Q$, which is given as

$$\Delta Q = C \times \Delta V$$

If we divide both sides by the change in time over which this took place, $\Delta t$, we find

$$\frac{\Delta Q}{\Delta t} = C \times \frac{\Delta V}{\Delta t}$$

But, $\Delta Q / \Delta t$ is the rate of flow of charge, i.e. the current, and so we have the result

$$I = C \times \frac{\Delta V}{\Delta t} \qquad (11.8)$$

Thus, when the switch is closed, the p.d. across the capacitor plates changes from its initial value of zero. Since the p.d. changes, the charge on each plate of the capacitor changes and, therefore, a current exists.

### Qualitative discussion

Let us give a brief discussion of what we can expect when we close the switch S in the circuit depicted in Figure 11.8. When the switch S is closed, the e.m.f. $E$ is applied abruptly to the circuit components. At the instant of closing the switch, the capacitor is uncharged and, therefore, there is no p.d. across it. It follows from Kirchhoff's voltage law that all the e.m.f. appears instantaneously across the resistor. Since the p.d. across the resistor, $V_R$, is equal to the e.m.f. of the battery, $E$, the maximum current flows at this time and is

$$I_{max} = \frac{V_R}{R} = \frac{E}{R}$$

This initial flow of current charges the capacitor (see Equation (11.8)) and we use the definition of current as the rate of flow of charge. If the capacitor is being charged, this must mean that the p.d. across the capacitor, $V_C$, must rise from its initial value of zero.

As $V_C$ increases then, according to Kirchhoff, the p.d. across the resistor falls, since we have at any instant

$$E = V_R + V_C$$

Hence, the charging current

$$I = \frac{V_R}{R} = \frac{E - V_C}{R}$$

falls from its maximum value.

Thus, as the p.d. across the capacitor rises, the charging current and, therefore, the *rate* at which the capacitor is charged will fall. The p.d. across the capacitor must eventually reach the e.m.f. $E$ of the battery. When this occurs the p.d. across the resistor is zero and it follows that the charging current goes to zero. If there is no more charging current, the charge on the capacitor remains constant and, therefore, the p.d. across the capacitor remains at the e.m.f. of the battery.

The relationship between the current, the charge stored by the capacitor and the p.d. across the capacitor is illustrated in Figure 11.9. Note that the p.d. across the capacitor is in a direction which inhibits further charging of the capacitor; hence, capacitance may be thought of as a property of a body which opposes voltage *changes* across itself.

### The solution

We quote the solution of the current at time $t$ after the switch has been closed as

$$I(t) = \frac{E}{R} \exp\left(-\frac{t}{RC}\right) \tag{11.9}$$

and from this the p.d. across the capacitor is

$$V_C(t) = E \times \left(1 - \exp\left(-\frac{t}{RC}\right)\right) \tag{11.10}$$

The exponential function is discussed in the Appendix.

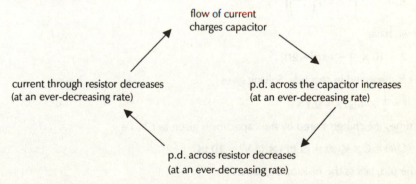

Figure 11.9   **The relationship between changing current and the value of the current**

We can write Equation (11.10) in a slightly different way by defining the 'time constant' by the equation

$$\tau = R \times C \tag{11.11}$$

so that we have

$$V_C(t) = E \times \left(1 - \exp\left(-\frac{t}{\tau}\right)\right) \tag{11.12}$$

That the capacitance is a measure of the opposition to voltage changes is reflected in the time constant. As the capacitance increases, so does the time constant and, therefore, the time taken for the capacitor to charge up to its maximum value (and also for the p.d. across the capacitor to change from its initial value of zero to $E$, the e.m.f. of the battery).

The variation of the p.d. across the capacitor with time is given in Figure 11.10, and is in agreement with that expected, i.e. a rapid rise from zero when the switch is first closed, to the smaller increase as the capacitor becomes charged and the final p.d. is the e.m.f. $E$ of the battery.

Worked Example 11.4 summarizes the above discussion.

━━━━ **WORKED EXAMPLE 11.4**

A battery of e.m.f. 10 V is connected to a 100 kΩ resistor in series with a 5 µF capacitor which is initially uncharged at time $t = 0$. How long does it take for the p.d. across the capacitor to reach a value of 2 V? At this time what is the charge stored by the capacitor, the p.d. across the resistor and the current flowing through the circuit?

− The time constant for the circuit is

$$\tau = R \times C = (100 \times 10^3 \ \Omega) \times (5 \times 10^{-6} \ F) = 0.5 \ s$$

To determine the p.d. across the capacitor at any time $t$ after the switch is closed we use the equation

$$V_C(t) = E \times \left(1 - \exp\left(-\frac{t}{\tau}\right)\right)$$

so that we have

$$2 = 10 \times (1 - \exp(-2t))$$

where $t$ is measured in seconds. Solving gives

$$t = \tfrac{1}{2} \times \ln(\tfrac{5}{4}) \ s = 0.1 \ s$$

At this time, the charge stored by the capacitor is given as $CV$, i.e.

$$Q(t) = C \times V_C(t) = (5 \ \mu F) \times (2 \ V) = 10 \ \mu C$$

while the p.d. across the resistor is given as

$$V_R = E - V_C = (10 - 2) \ V = 8 \ V$$

We can use the formula given above for the current through the circuit at any time *t*. However, it is more convenient to make use of Ohm's law, since we know the p.d. across the resistor. We find

$$I(t) = \frac{V_R}{R} = \frac{8\,V}{100\,k\Omega} = 80\,\mu A$$

## Analysis based on Kirchhoff's laws

We now provide a technical discussion based upon Kirchhoff's laws, see Figure 11.11.

Suppose at time *t* after the switch has been closed, the current through the circuit is *I*, the p.d. across the resistor is $V_R$ and the p.d. across the capacitor is $V_C$. Then, according to Kirchhoff's voltage law, we have

$$E = IR + V_C$$

where we have used the usual relation between the current through and p.d. across the resistor. If the magnitude of the charge stored on each plate at this time is *Q*, then we can write $V_C$ in terms of *Q* and the capacitance *C*; see (11.2) or (11.3). We find

$$E = IR + \frac{Q}{C}$$

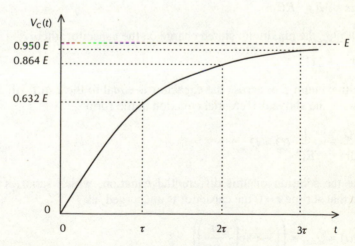

*Figure 11.10* **The variation of the p.d. across the capacitor with time**

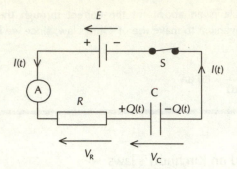

*Figure 11.11* **Capacitance in a circuit and Kirchhoff's laws**

To proceed we have to use that the current, $I$, is the rate of flow of charge (see Equation (3.1)), i.e.

$$I = \frac{dQ}{dt}$$

Note that we have to make full use of calculus notation and have replaced

$$\frac{\Delta Q}{\Delta t} \rightarrow \frac{dQ}{dt}$$

Thus, we have derived the following differential equation

$$\frac{dQ}{dt} = \frac{E}{R} - \frac{Q}{RC}$$

If we call $Q_{max}$ the maximum stored charge on the capacitor, then

$$Q_{max} = EC$$

since the maximum p.d. across the capacitor is equal to the e.m.f. of the battery, $E$. We can write the above differential equation in the form

$$\frac{dQ}{dt} = -\frac{1}{RC}(Q - Q_{max})$$

We quote the solution of this differential equation, which satisfies the boundary condition that at time $t = 0$ the capacitor is uncharged, as

$$Q(t) = Q_{max} \times \left(1 - \exp\left(-\frac{t}{RC}\right)\right)$$

From this result we can derive the following

$$V_C(t) = E \times \left(1 - \exp\left(-\frac{t}{RC}\right)\right)$$

and

$$I(t) = \frac{dQ}{dt} = \frac{E}{R} \exp\left(-\frac{t}{RC}\right)$$

In this result, notice that $E/R$ is the current that would flow if the capacitor were replaced by a conductor.

## 11.5 Discharging a capacitor

Suppose a capacitor has stored a charge $Q_{max}$, then the p.d. across the capacitor is given as

$$V_C = \frac{Q_{max}}{C}$$

If we connect the capacitor across its terminals with a resistor, the capacitor will discharge. This follows since there is a charge imbalance between the capacitor plates, i.e. one plate has charge $+Q_{max}$ while the other plate has charge $-Q_{max}$. The charge will flow until this imbalance is removed; if charge moves this implies that a current will exist. It is of interest to determine the time dependence of this discharge current.

Before proceeding with the analysis, we note that we expect the current in the circuit depicted in Figure 11.12 to flow from the plate charged to $+Q$ to the plate charged with $-Q$. Why have we drawn the current in the figure flowing in the

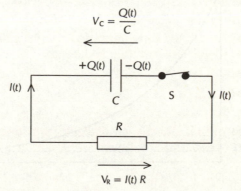

*Figure 11.12* **The discharge of a capacitor**

opposite direction? This is a technical matter relating to calculus, where the current is strictly the rate of *increase* of charge flowing. Placing the charges $+Q$ and $-Q$ as we have in Figure 11.12 means that we have no choice in the direction of the current flow.

We simply quote the solution for the p.d. across the capacitor at time $t$ after the switch has been closed as

$$V_C(t) = V_{max} \exp\left(-\frac{t}{\tau}\right)$$

where $V_{max}$, the initial p.d. across the capacitor, is

$$V_{max} = \frac{Q_{max}}{C}$$

and $\tau$ is the time constant as defined in (11.11).

Finally, the current flowing at any time $t$ is given as

$$I(t) = -\frac{V_{max}}{R} \exp\left(-\frac{t}{RC}\right)$$

the negative sign indicating that the current is flowing in the opposite direction to that assumed, as expected. Figure 11.13 shows the time dependence of the discharge.

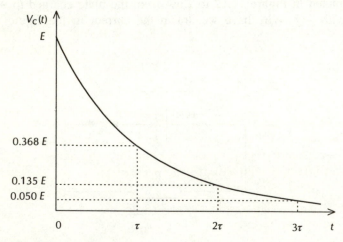

*Figure 11.13*   **The discharge of a capacitor**

○     ## Analysis based on Kirchhoff's laws

From Kirchhoff's laws we have

$$V_C + V_R = 0$$

from which we can write

$$\frac{dQ}{dt} = -\frac{1}{RC} Q$$

This simple first-order differential can be solved easily. We find that the charge stored by the capacitor as a function of the time $t$ after the switch S has been closed is

$$Q(t) = Q_{max} \exp\left(-\frac{t}{RC}\right)$$

It follows that the p.d. across the capacitor at time $t$ is

$$V_C(t) = V_{max} \exp\left(-\frac{t}{RC}\right)$$

where $V_{max}$ is the initial p.d. across the capacitor and is

$$V_{max} = \frac{Q_{max}}{C}$$

## 11.6  Energy storage

Suppose that a capacitor has stored a charge $Q$. If $C$ is the capacitance, the potential difference across the plates is

$$V = \frac{Q}{C}$$

Since like charges repel, when we bring an extra amount of charge $\Delta Q$ to the capacitor we have to do work. If the amount of charge $\Delta Q$ is so small that the p.d. between the capacitor plates remains constant the extra amount of work we have to do to move this charge onto the capacitor is given as

$$\Delta W = (\text{p.d.}) \times (\text{charge moved}) = \frac{Q}{C} \times \Delta Q$$

This follows from the definition of the volt.

To charge a capacitor from $Q = 0$ to final charge $Q$ we use an amount of work

$$W = \frac{1}{2} \frac{Q^2}{C}$$

Using the definition of capacitance in Equations (11.2) and (11.3), we can write this in the form

$$W = \tfrac{1}{2} C V^2$$

Thus, when we charge up a capacitor, it stores an amount of energy equal to the work done in charging the capacitor. This energy is said to reside in the electric field between the plates.

## 11.7  Summary

We described the action of a pair of conducting plates connected to a battery and argued that the plates act so as to store electric charge. We referred to this property as capacitance.

After formally defining the capacitance of a body in terms of the rise in the p.d. of the body when charge is acquired, we then went on to discuss the case of the parallel plate capacitor and some practical aspects of capacitors. The effective capacitances of capacitors connected in series and in parallel were also discussed.

We presented an argument which suggested that when a capacitor is connected across a resistor and battery that capacitance is a measure of the opposition to voltage changes. The cases of a capacitor charging and discharging were discussed.

It was noted that as a capacitor is being charged it is necessary to do work to move like charges close to each other. It follows that when charged, the capacitor stores energy. This energy is said to reside in the electric field between the plates.

## ■■■ Problems

In the following questions take $\varepsilon_0$ as $8.854 \times 10^{-12}\ \mathrm{F\,m^{-1}}$.

**11.1.**  How long does it take a 1 µF capacitor to charge to 10 V with a constant current of 1 mA?

**11.2**  What current is needed to uniformly charge a 10 pF capacitor to 5 V in 4 ns?

**11.3**  What capacitance will charge to 10 V in 50 µs with a constant current of 10 mA?

**11.4**  Determine the capacitance of a parallel plate capacitor with the dimensions $A = 100\ \mathrm{mm^2}$ and $d = 0.001$ mm, taking the plates to be separated by free space.

(a)  A battery of e.m.f. 10 V is connected across the plates. The capacitor is fully charged and then the battery is removed. Determine the charge stored on each plate.

(b)  If the common area of the plates is reduced to 5 mm², determine the new p.d. across the capacitor.

(c)  Determine the energy stored by the capacitor in parts (a) and (b) and comment on your result.

**11.5**   Two capacitors $C_1 = 10\ \mu F$ and $C_2$ are placed in parallel and have a capacitance $C = 35\ \mu F$. Determine the value of $C_2$.

**11.6**   Two capacitors $C_1 = 25\ \mu F$ and $C_2 = 125\ \mu F$ are placed in series. Determine the effective capacitance. The capacitors are originally uncharged and then connected to a 10 V battery. Once they are fully charged, determine the charge on each capacitor, the p.d. across each capacitor and the energy stored by each capacitor.

**11.7**   Determine the capacitance for the arrangement shown in Figure P11.1. The common area of the plates is $A = 10^{-4}\ m^2$ and their separation is $d = 1\ \mu m$. The material is placed halfway between the plates and has relative permittivity of 5. Take air to have relative permittivity of 1.0.

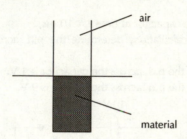

*Figure P11.1*

*Hint:* Treat as two capacitors in parallel, each of cross-sectional area equal to $A/2$.

**11.8**   Determine the capacitance of the capacitor network shown in Figure P11.2. If a battery of e.m.f. 20 V is connected across the capacitor network determine the p.d. across and the magnitude of the charge on each capacitor.

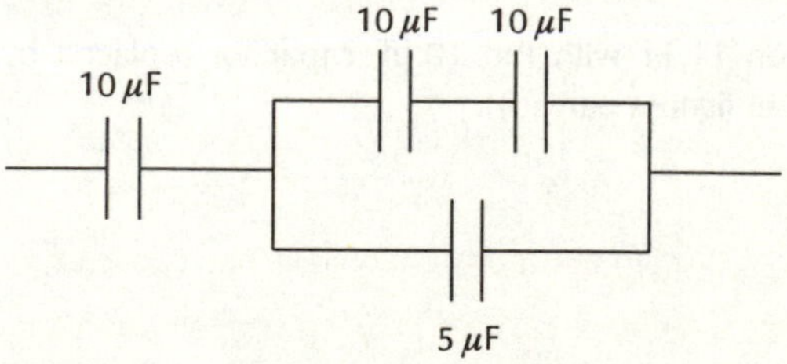

*Figure P11.2*

**11.9**   Determine the current during the times AB, CD and EF when the p.d. across a 5 μF capacitor depends on time, as indicated in Figure P11.3.

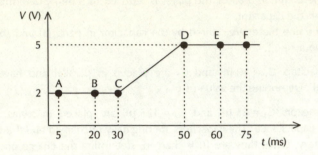

*Figure P11.3*

○ **11.10** Show that *RC* has the units of time.
Hint: Make use of the results of Chapter 1.

**11.11** Consider the circuit shown in Figure P11.4. If the switch is closed at time $t = 0$, determine
(a) the time constant,
(b) the p.d. across the capacitor at time $t = 10$ ms,
(c) *without detailed calculation*, determine the p.d. across the resistor at the same time,
(d) the time at which the p.d. across the capacitor is 1 V.
(e) the time at which the p.d. across the resistor is 9 V.

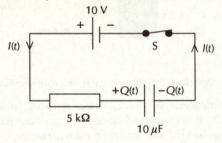

*Figure P11.4*

**11.12** Repeat Question 11.11 with the 10 μF capacitor replaced by two 20 μF capacitors placed in parallel (ignore part (d)).

# The responses of resistors, inductors and capacitors to an alternating current

In Chapters 10 and 11 we met the concept that the current through a circuit can depend upon the time. In Chapter 10 we discussed the time dependence of the current through the circuit consisting of a resistor connected in series with an inductor to a source of *constant* e.m.f., i.e. a battery. Similarly, in Chapter 11 we discussed the time dependence of a current through the circuit consisting of a resistor and inductor and a resistor and capacitor connected in series to a battery. In both cases the currents are referred to as 'DC transients'. The term 'DC' stands for 'direct current' and refers to the fact that the current does not change direction through the circuit, while the term 'transient' implies that the time dependence of the current eventually falls off. In the case of the resistor–inductor circuit we found that the current starts off from zero and eventually reached the *steady* value given by Ohm's law, while for the resistor–capacitor circuit the current started off at the Ohm's law value and eventually decayed to zero.

In this chapter we will discuss more examples where the current through a circuit depends upon time. Here, however, we consider circuits where the behaviour is different from that described previously in two major ways:

■ The current changes direction, i.e. for some time the current flows in one direction through the circuit and later flows in the opposite direction.
■ This behaviour continues for as long as the source of e.m.f. driving the current through the circuit is connected, i.e. it is not a transient behaviour.

We begin by recalling previous work relating to alternating voltage and current supplies. We emphasize a particular example of a wave form – namely, the sinusoidal wave form – and present several definitions and properties of such a wave form. It will then be useful to consider two sinusoidal waves which have the same time dependence (i.e. the same frequency) but which are out of step with each other. The amount of this 'out-of-step-ness' is called the *phase shift*.

We proceed further by discussing the behaviour of circuits consisting of a resistor, an inductor and a capacitor separately connected to an alternating voltage supply. We will refer to circuits connected to a sinusoidal voltage source as AC circuits, where 'AC' stands for *alternating current*. This is a reasonable notation since we will find that the current through such a circuit is constantly varying in both size and direction as the time increases. The reason why the current constantly changes direction is straightforward: the polarity of the e.m.f. driving the current through a circuit is changing with time.

For a sinusoidal voltage supply connected across a component we shall find that the current has the same frequency as the supply – this is perhaps an obvious conclusion, since it merely tells us that the e.m.f. changes . However, we shall note that the current wave form will be shifted or out of step with respect to the voltage wave form by an amount given by the phase shift.

In this chapter we consider only circuits consisting of either a single resistor, inductor or capacitor connected to an AC supply. These circuits will be analyzed analytically by making a simple application of Kirchhoff's laws.

It will then be convenient to introduce phasors and the phasor diagram. Phasors are a geometrical way of representing time-varying quantities, such as sine waves in terms of their magnitude and phase. We will then make a link between the preceding work and phasors. The phasor approach will be very useful when we consider AC circuits made up of two or more components.

## 12.1   The sinusoidal wave form

In Section 9.3 we noted that a coil rotating in a magnetic field generates an e.m.f. which depends upon time in the manner indicated by Equation (9.7). We proceed further by writing (9.7) in the more general form

$$\mathcal{E}(t) = \mathcal{E}_0 \sin(\omega t) \tag{12.1}$$

This *sinusoidal wave form* is depicted in Figure 12.1, which shows several complete *cycles* of the variation of the induced e.m.f. as the time varies. This figure introduces some very important notation.

From the properties of the sine function we have that the induced e.m.f. varies between $+\mathcal{E}_0$ and $-\mathcal{E}_0$, where the value $\mathcal{E}_0$ is referred to as the *amplitude* or the *peak value* of the wave form. The voltage measured between the positive and negative peaks, $2\mathcal{E}_0$, is referred to as the *peak-to-peak value*.

A cycle is defined as the interval between any two corresponding points on the wave form. The time for one complete cycle is called the *period, T*. From the properties of the sine function there is an important relationship between the angular frequency $\omega$ and the period $T$ which we now derive.

At some particular time $t$ the value of the e.m.f. is given by (12.1) as

$$\mathcal{E}(t) = \mathcal{E}_0 \sin(\omega t)$$

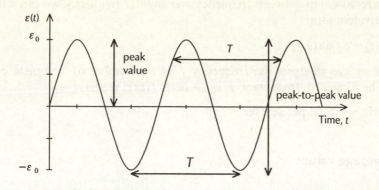

*Figure 12.1* **The sinusoid waveform**

At the later time $t + T$ we have

$$\mathscr{E}(t + T) = \mathscr{E}_0 \sin(\omega(t + T))$$

Expanding the right-hand side of the above gives $\mathscr{E}(t + T)$ as

$$\mathscr{E}_0 \sin(\omega t + \omega T)$$

We have defined the period, $T$, as the time for which the wave form takes the same value, so that we must also have

$$\mathscr{E}(t + T) = \mathscr{E}(t)$$

We now make use of the fact that the sine function is a periodic function with the property that, for any value of the angle $\theta$

$$\sin(\theta + 2\pi) = \sin(\theta)$$

i.e. it repeats itself every $2\pi$ radians. The sinusoidal wave form is a very important example of a *periodic* wave form; the term 'periodic' simply means that the wave form repeats itself after the angle $\theta$ has increased by some value. The Appendix contains some useful details on trigonometric functions.

From this we must have that $\omega T$ is equal to $2\pi$, and it therefore follows that the period, $T$, is related to the angular frequency of the supply, $\omega$, as

$$T = \frac{2\pi}{\omega}$$

It is convenient to introduce the *frequency, f*, related to the angular frequency $\omega$ by the equation

$$\omega = 2\pi f$$

so that the relationship between the period and the frequency is given as

$$T = \frac{1}{f}$$

Given the relationship between frequency and angular frequency we can write (12.1) in the equivalent form

$$\mathcal{E}(t) = \mathcal{E}_0 \sin(2\pi f t) \tag{12.2}$$

Note that we can interpret the frequency $f$ as the number of complete cycles per second. The SI unit of frequency, $f$, is the hertz (Hz). We have

$$1 \text{ Hz} = 1 \text{ cycle per second}$$

## 12.2 Average values

As mentioned previously, the magnitude and the polarity of the e.m.f. depend upon time. It is obviously very inconvenient to record all the values of the e.m.f. for all times. Rather, we find it very useful to have just a single number which represents some average property of the wave form. We have previously met some such numbers, namely, the peak value and the peak-to-peak value. However, these just give information on the range of the e.m.f. What would be more useful is a number which gives some 'average' or 'typical value' of the sinusoidal supply.

### Introduction to the meaning of 'average value'

Before discussing circuits proper we first introduce what we mean by the term 'average value'. We begin by assuming we have some function which depends upon time in the manner indicated in Figure 12.2. We denote the function $f$ by $f(t)$, where this just means that the function $f$ depends upon time. As a definite example we will take the function $f(t)$ to represent the speed of a particle.

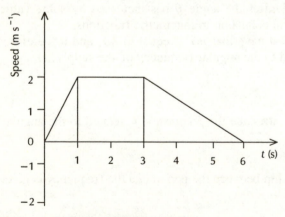

*Figure 12.2*   **A particle moving with different speeds**

With this example, the figure depicts a particle accelerating from rest between $t = 0$ and $t = 1$ s, and then moving at a constant speed of $2$ m s$^{-1}$ for the next two seconds. For the next four seconds the particle decelerates, until it becomes stationary (i.e. its speed is zero).

Suppose we wanted to determine the average speed of the particle (in the following we will refer to the particle as the 'real particle'). What does the term 'average speed' mean? To explain this, we imagine a second particle which starts off at the same time and position as our real particle, but moves with a constant speed, call it $v_{ave}$. The speed $v_{ave}$ is chosen such that the second particle covers the same distance in the same time as the real particle. Since the second particle moves with a constant speed, after six seconds it has travelled a distance.

distance travelled = speed × time = $v_{ave} \times 6$ s

What is the distance travelled by the first particle? It is given by the area under the curve of speed versus time. The area is determined in Figure 12.3. For reasons that will become apparent soon, we also show in Figure 12.3 how the distance travelled by the second particle is given in terms of the area under its speed versus time curve.

Collecting our results, we have

distance travelled = 8 m

$$\Rightarrow v_{ave} = \frac{8 \text{ m}}{6 \text{ s}} = 1.333 \text{ m s}^{-1}$$

We generalize the above result and state that if $\mathscr{F}(t)$ is some given function of time, the average of this quantity over the times $t_1$ to $t_2$ is *defined* as

$$\langle \mathscr{F} \rangle = \frac{1}{t_2 - t_1} \int_{t = t_1}^{t = t_2} \mathscr{F}(t) \, dt \qquad (12.3)$$

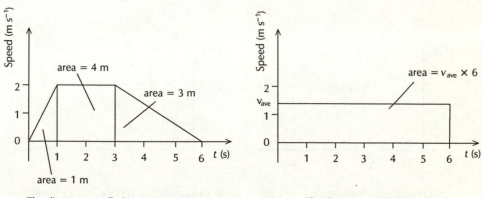

The distance travelled is $(1 + 4 + 3) = 8$ m          The distance travelled is $6v_{ave}$

*Figure 12.3* **The area under the curve is the distance travelled**

This is easily interpreted: the integral of $\mathcal{F}(t)$ over the times $t_1 \leq t \leq t_2$ just gives the area of the curve $\mathcal{F}(t)$ between these two times. The average value of the quantity $\mathcal{F}$, denoted by the angle brackets $\langle \mathcal{F} \rangle$, is just the area of $\mathcal{F}$ between the times $t_1$ and $t_2$, divided by the length of the limits of the integral. Note that we can write (12.3) in the following form:

$$\int_{t=t_1}^{t=t_2} \mathcal{F}(t)\, \mathrm{d}t = \langle \mathcal{F} \rangle \times (t_2 - t_1)$$

We can then interpret the average value, $\langle \mathcal{F} \rangle$, such that $\langle \mathcal{F} \rangle \times (t_2 - t_1)$ is equal to the area under the curve $\mathcal{F}(t)$ between the limits $t_1$ and $t_2$; see Figure 12.4. Hence, the definition (12.3) represents exactly what we did in the example. We have just gone one (mathematical) step further by relating 'area under curve' to the integral operation.

### The average value of a sinusoidal wave form

Figure 12.5 shows one cycle of the sinusoidal supply. From the definition (12.3) it is straightforward to determine that the average value of the sinusoidal supply, $\mathcal{E}(t)$, over one complete cycle is zero, i.e.

$$\langle \mathcal{E} \rangle_{\text{one cycle}} = \frac{1}{T} \int_{t=0}^{t=T} \mathcal{E}_0 \sin(2\pi f t)\, \mathrm{d}t = 0 \qquad (12.4)$$

where we have made use of the relationship between the period, $T$, and the frequency, $f$.

Note that it is easy to interpret this result. It is apparent that the area under the curve in the first half of the cycle is equal in magnitude but opposite in sign to the area under the curve in the second half of the cycle. Thus, the total area under the curve is zero.

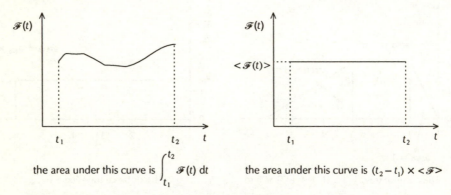

Figure 12.4   **Defining the average value of a function**

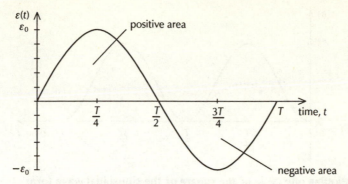

*Figure 12.5*    **The average value of one cycle of the sinusoidal wave form is zero**

It follows that this average value is not of much use. One way to overcome this problem is to define an average value over just half a cycle. In this case the supply is at all times the same sign so that it is not possible for the integral in (12.3) to vanish. If we do this we find

$$\langle \mathscr{E} \rangle_{\text{half cycle}} = \frac{1}{T/2} \int_{t=0}^{t=T/2} \mathscr{E}_0 \sin(2\pi f t) \, dt = \frac{2}{\pi} \mathscr{E}_0 \tag{12.5}$$

### The root mean square value

We noted that the reason why the average value given by (12.4) was zero was that the area under the curve of the supply over one complete cycle was zero. This led us to consider a half cycle, since in this case the value of the supply is at all times the same sign. Another method is to consider the average value of the square of the supply over a cycle. By considering the square we note that the area will necessarily come out as non-zero; see Figure 12.6. Note that in Figure 12.6 the dashed curve represents the original sinusoidal wave form for comparison. We have

$$\langle \mathscr{E}^2 \rangle_{\text{m.s.}} = \frac{1}{T} \int_{t=0}^{t=T} \mathscr{E}_0^2 \sin^2(2\pi f t) \, dt = \frac{1}{2} \mathscr{E}_0^2 \tag{12.6a}$$

which is the average of the square of the sinusoidal supply over one cycle, and is referred to as the *mean square (m.s.) value*.

The *root mean square (r.m.s.) value*, $\mathscr{E}_{\text{r.m.s.}}$ is the square root of the mean square value, i.e.

$$\mathscr{E}\text{r.m.s.} = \sqrt{\langle \mathscr{E}^2 \rangle_{\text{m.s.}}}$$

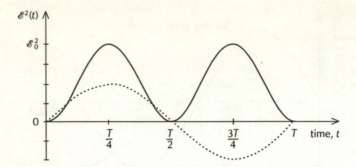

*Figure 12.6* **Plotting one cycle of the square of the sinusoidal wave form**

In our particular example we have

$$\mathscr{E}_{\text{r.m.s.}} = \frac{1}{\sqrt{2}} \mathscr{E}_0 \tag{12.6b}$$

We will discuss later why this is a useful measure.

Worked Example 12.1 brings together the concepts discussed above.

━━━━ **WORKED EXAMPLE 12.1**
───────────────────────────────────────────────

A sinusoidal wave form has an r.m.s. value of 14.14 V and a frequency of 50 Hz. Determine the peak to peak value of the source, the angular frequency of the source and the time for which the source repeats itself.

− The relationship between the peak value and the r.m.s. value (Equation (12.6 b)) gives

$$\mathscr{E}_{\text{peak}} = \mathscr{E}_0 = \sqrt{2}\,\mathscr{E}_{\text{r.m.s.}}$$

from which we find

$$\mathscr{E}_{\text{peak}} = \sqrt{2} \times 14.14\ \text{V} = 20\ \text{V}$$

and, hence,

$$\mathscr{E}_{\text{peak-to-peak}} = 2 \times \mathscr{E}_{\text{peak}} = 40\ \text{V}$$

Given the frequency as $f = 50$ Hz, the angular frequency $\omega$ and the period $T$ are given as

$$\omega = 2\pi f = 314\ \text{rad s}^{-1}$$

and

$$T = \frac{1}{f} = \frac{1}{50}\ \text{s} = 20\ \text{ms}$$

i.e. the time taken for one complete cycle is 20 ms.

────────────────────────────────────────────────────────

Worked Example 12.2 considers other wave forms than the sinusoidal wave form.

━━━ **WORKED EXAMPLE 12.2**

For each of the periodic wave forms shown in Figure 12.7, determine the frequency, the average value over half a cycle, and the r.m.s. value.

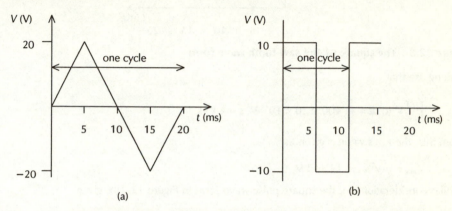

*Figure 12.7* **(a) The saw-tooth wave form; (b) the square pulse wave form**

− For the saw-tooth wave form depicted in Figure 12.7(a) we have that the wave form repeats itself every 20 ms. Therefore, the period $T$, is known. It follows that the frequency is given as

$$f = \frac{1}{T} = \frac{1}{20 \times 10^{-3}\,\text{s}} = 50\,\text{Hz}$$

The average value over half a cycle is determined from (12.3), i.e.

$$\langle V \rangle = \frac{1}{T/2} \int_{t=0}^{t=T/2} V(t)\, dt$$

But

$$\int_{t=0}^{t=T/2} V(t)\, dt = \frac{1}{2}\,(20\,\text{V}) \times (10 \times 10^{-3}\,\text{s}) = 0.1\,\text{V s}$$

since this is just the area under the triangle. It therefore follows that the average value over half a cycle is given as

$$\langle V \rangle_{\text{half cycle}} = \frac{1}{10 \times 10^{-3}\,\text{s}} \times 0.1\,\text{V s} = 10\,\text{V}$$

The r.m.s. square value is given as the square root of the mean of the square of the wave form. The square of the saw-tooth wave form is given in Figure 12.8 and is again a saw-tooth-like wave form. This gives the mean square value as

$$\langle V^2 \rangle_{\text{m.s.}} = \frac{1}{T} \int_{t=0}^{t=T} V^2(t)\, dt = 200\,\text{V}^2$$

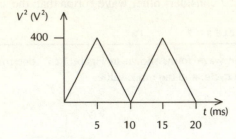

*Figure 12.8* **The square of the saw-tooth wave form**

making use that

$$\int_{t=0}^{t=T} V^2(t)\, dt = \frac{1}{2}\, 400 \times 20 \times 10^{-3}\ V^2\ s = 4\ V^2\ s$$

From this, the r.m.s value is given as

$$V_{r.m.s} = \sqrt{\langle V^2 \rangle_{m.s.}} = 14.142\ V$$

Similar considerations for the square pulse wave form in Figure 12.7(b) gives

$$f = \frac{1}{T} = \frac{1}{10 \times 10^{-3}\ s} = 100\ Hz$$

$$\langle V \rangle_{\text{half cycle}} = 5\ V$$

$$V_{r.m.s.} = 7.071\ V$$

## 12.3 The phase shift

The concept of 'phase shift' will be very important in the ensuing work, and it is convenient to discuss it here. To introduce the concept of the phase shift it is necessary to consider the following sinusoidal wave forms:

$$\mathscr{E}_2(t) = \mathscr{E}_0 \sin(2\pi ft + \phi) \tag{12.7a}$$
$$\mathscr{E}_3(t) = \mathscr{E}_0 \sin(2\pi ft - \phi) \tag{12.7b}$$

The wave forms (12.7a) and (12.7b) both vary with the same frequency, $f$, as the sinusoidal wave form $\mathscr{E}(t)$ given by Equation (12.2):

$$\mathscr{E}(t) = \mathscr{E}_0 \sin(2\pi ft) \tag{12.2}$$

The difference between Equations (12.7) and (12.2) is that in (12.7) they both have an extra factor, $\phi$, called the phase shift. The significance of the phase shift is indicated in Figures 12.9(a) and (b).

Figure 12.9(a) shows the case of wave forms (12.7a) and (12.2). At time $t = 0$, note that $\mathscr{E}(t)$ has the value zero, while $\mathscr{E}_2(t)$ has the value $\mathscr{E}_0 \sin(\phi)$. Also the wave

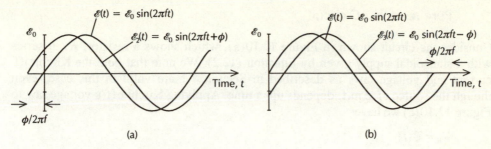

*Figure 12.9* **Introducing the phase shift**

form $\mathscr{E}_2(t)$ reaches its maximum value some time before $\mathscr{E}(t)$ reaches its maximum value. The wave forms $\mathscr{E}(t)$ and $\mathscr{E}_2(t)$ have the same shape, except that $\mathscr{E}_2(t)$ has been translated to the left compared to the wave form $\mathscr{E}(t)$. We say that the wave form $\mathscr{E}_2(t)$ *leads* $\mathscr{E}(t)$ by the phase shift $\phi$.

Figure 12.9(b) shows the case of the wave forms (12.7b) and (12.2). Again, the wave form $\mathscr{E}_3(t)$ is the same shape as that given by $\mathscr{E}(t)$, except that it has now been translated to the right compared to the wave form $\mathscr{E}(t)$. We say that the wave form $\mathscr{E}_3(t)$ *lags* $\mathscr{E}(t)$ by the phase shift $\phi$.

## 12.4 AC and R, L and C circuits

We now wish to discuss the *frequency response* of various circuits connected to a sinusoidal supply. The frequency response of the circuit gives us information about the relationship between the p.d. across the component and the current through it. In particular, we are interested in the amplitude of the current through the circuit and by how much it is out of step with respect to the source e.m.f. (i.e. the *phase shift*), as well as how these parameters depend upon the frequency of the supply.

In particular, we discuss the frequency response of a circuit consisting of a single resistor, $R$, an inductor, $L$, and a capacitor, $C$, connected to a sinusoidal supply, as shown in Figures 12.10(a), (b) and (c), respectively.

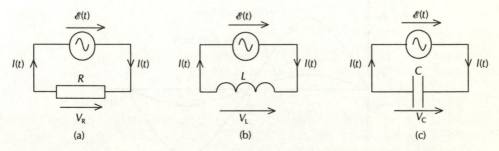

*Figure 12.10* **The effects of a sinusoidal supply on pure resistive, inductive and capacitive circuits**

## Pure resistive AC circuit

Consider the circuit shown in Figure 12.10(a), which shows a resistor, $R$, in series with a sinusoidal supply given by Equation (12.2). We note that both the Kirchhoff current and voltage laws as described in Section 7.1 are valid in this case, even though the source of e.m.f. depends upon time. Applying Kirchhoff's voltage law to Figure 12.10(a) we have

$$V_R = \mathscr{E}(t)$$

i.e. at each instant, the p.d. across the resistor, $V_R$, is equal to the e.m.f. of the source, $\mathscr{E}(t)$. If the current through the resistor at any instant is $I$ then, according to Ohm's law, we have

$$I(t) = \frac{V_R}{R} = \frac{\mathscr{E}(t)}{R}$$

and so at each instant the current through the resistor, $I(t)$, is proportional to the p.d. across it, $V_R(t)$. Technically, we say the current through the circuit is *in phase* with the source of the e.m.f., $\mathscr{E}(t)$. For the case of $\mathscr{E}(t)$ given as the sinusoidal wave form (12.2) we have

$$I(t) = I_0 \sin(2\pi ft)$$

where $I_0$ is simply determined.

The wave form for the current through and the p.d. across the resistor is shown in Figure 12.11.

*Power in a pure resistive AC circuit*

The instantaneous power developed across the resistor is given by (6.5). We have

$$P(t) = I(t) \times V_R(t)$$

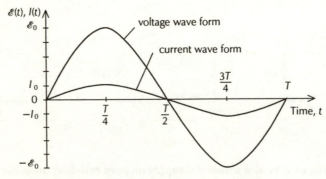

*Figure 12.11*   **The voltage and current wave forms for a resistor**

Figure 12.12 shows the instantaneous power dissipated by the resistor over one complete cycle. Since the current through the resistor is always in phase with the p.d. across the resistor, the instantaneous power dissipated by the resistor is always positive.

It is of interest to calculate the average power dissipated by the resistor over one cycle, $P_{ave}$. We have

$$P_{ave} = \langle P \rangle_{\text{one cycle}} = \frac{1}{T} \int_{t=0}^{t=T} \mathscr{E}_0 I_0 \sin^2(2\pi f t)\, dt = \frac{1}{2} I_0 \mathscr{E}_0$$

See the Appendix for useful details.

We note that this result can be written in terms of the r.m.s. values of the p.d. and the current as

$$P_{ave} = I_{r.m.s.} \mathscr{E}_{r.m.s.}$$

where we have used (12.6b) and a similar result for the r.m.s. value of the current.

### The r.m.s. value of an AC and the equivalent DC

Note that we can write the power dissipated by the resistor in the form

$$P_{ave} = I_{r.m.s.}^2 R$$

Now the heating effect of an electric current is independent of the direction of the current flow through a resistor so we can interpret the current $I_{r.m.s.}$ as that value of the DC current which would produce the same heating effect as the time varying AC through the resistor.

### Pure inductive AC circuit

Refer to the circuit shown in Figure 12.10(b), which shows an inductor, $L$, in series

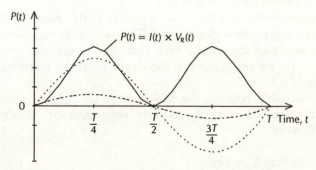

*Figure 12.12* **The power developed across a resistor is always positive**

with a sinusoidal supply given by (12.2). Again we apply Kirchhoff's voltage law, from which we find

$$V_L = \mathscr{E}(t)$$

According to previous results (see Section 10.3 and (10.11) in particular), we can write

$$L \frac{dI}{dt} = \mathscr{E}(t)$$

Note that this equation relates the *rate of change* of the current with respect to time to the e.m.f., and not the current directly. This is a very important observation. For the case of $\mathscr{E}(t)$ the sinusoidal wave form (12.2) we have

$$L \frac{dI}{dt} = \mathscr{E}_0 \sin(2\pi f t)$$

The solution of this is straightforward, and we find that the current through the circuit is

$$I(t) = -\frac{\mathscr{E}_0}{2\pi f L} \cos(2\pi f t)$$

Note that we can make use of well known properties of the sine and cosine functions (see the Appendix) to write the above in the form

$$I(t) = I_0 \sin\left(2\pi f t - \frac{\pi}{2}\right)$$

where we have also introduced the maximum or peak current through the circuit, $I_0$, which is related to $\mathscr{E}_0$, $L$ and $f$ as

$$I_0 = \frac{\mathscr{E}_0}{2\pi f L}$$

There are two important points to note here:

*(1)*   The current varies with the same frequency as the source of the e.m.f., but it is no longer in step or phase with the e.m.f. The current through the circuit is *out of phase* with the source of the e.m.f., $\mathscr{E}(t)$. Here the current *lags* by $\pi/2$ behind the e.m.f. Alternatively, we can argue that the e.m.f. *leads* the current by $\pi/2$. This is shown in Figure 12.13.

*(2)*   The relationship between the peak current through the circuit, $I_0$, and the peak p.d. across the inductor, $\mathscr{E}_0$, is very suggestive of an Ohm's law type relationship. We write

$$I_0 = \frac{\mathscr{E}_0}{X_L}, \quad \text{where } X_L = 2\pi f L \tag{12.8}$$

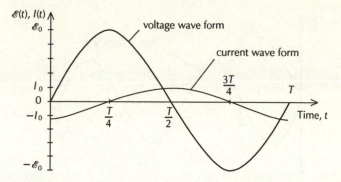

*Figure 12.13*  **The voltage and current wave forms for an inductor**

The quantity $X_L$ seems to be a resistance-like property of the circuit and is referred to as the inductive reactance. Note that the reactance depends linearly on frequency; see Figure 12.14.

That the reactance increases with frequency is reasonable – this follows since we previously suggested that inductance is a measure of the opposition to the *changes* in current. Thus, as the frequency of the supply increases, so the changes in the current increase, and therefore the larger the back e.m.f. induced in the inductor. In the limit as the frequency of the supply approaches zero the smaller the reactance becomes. In the limit of a zero frequency source the reactance is zero and the inductor acts like a short circuit. At the opposite extreme, for very large frequencies the inductor acts like an open circuit.

### Power in a pure inductive AC circuit

The instantaneous power developed across the inductor is given as the product of

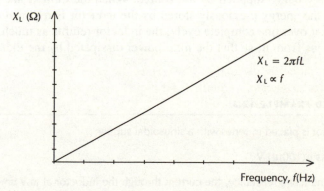

*Figure 12.14*  **The inductive reactance of an inductor versus the frequency**

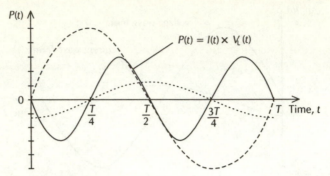

*Figure 12.15* **The power developed across an inductor over one complete cycle. Note that the power is no longer always positive**

the p.d. across the inductor and the current through it (recall Equation (6.5)). We have

$$P(t) = I(t) \times V_L(t)$$

Figure 12.15 shows how the instantaneous power dissipated by the inductor varies with time over one complete cycle. Since the current through the inductor is out of phase with the p.d. across the inductor the instantaneous power dissipated by the inductor is sometimes positive and sometimes negative.

It is of interest to calculate the average power dissipated by the inductor over one cycle, $P_{ave}$. We have

$$P_{ave} = \langle P \rangle_{\text{one cycle}} = \frac{1}{T} \int_{t=0}^{t=T} \mathcal{E}_0 I_0 \sin(2\pi ft)\sin\left(2\pi ft - \frac{\pi}{2}\right) dt = 0$$

Thus a perfect inductor dissipates no power. This is interpreted as follows: When the current and p.d. have the same sign, the inductor is storing energy in the magnetic field, this energy being supplied by the source. When the current and the p.d. have opposite signs the energy previously stored by the inductor is returned to the source. This implies that over one complete cycle, the inductor returns as much energy to the source as it stores from it, so that the *total* power dissipated by the inductor over one cycle is zero.

━━━ **WORKED EXAMPLE 12.3**

A 10 mH inductor is placed in series with a sinusoidal supply:

$$\mathcal{E}(t) = 5 \sin(100\pi t) \text{ V}$$

Determine the inductive reactance, the current through the inductor at any time and the r.m.s current.

– The inductive reactance is given as

$$X_L = 2\pi f L = 2 \times 3.142 \times (50 \text{ Hz}) \times (10 \times 10^{-3} \text{ H}) = 3.142 \ \Omega$$

since the frequency of the supply is 50 Hz.

The peak voltage of the supply is 5 V. Therefore, from (12.8), the peak current through the circuit is

$$I_0 = \frac{\mathscr{E}_0}{X_L} = \frac{5 \text{ V}}{3.142 \ \Omega} = 1.592 \text{ A}$$

It follows that the current through the circuit at any time is

$$I(t) = 1.592 \sin\left(100\pi t - \frac{\pi}{2}\right) \text{ A}$$

since the current lags by $\pi/2$ behind the e.m.f.

The r.m.s. current is given as

$$I_{r.m.s.} = \frac{1}{\sqrt{2}} I_0 = 1.126 \text{ A}$$

---

## Pure capacitive AC circuit

Refer to the circuit shown in Figure 12.10(c), which shows a capacitor, $C$, in series with a sinusoidal supply given by (12.2). Again we apply Kirchhoff's voltage law, from which we find

$$V_C = \mathscr{E}(t)$$

According to previous results (see Section 11.1 and Equation (11.3) in particular), we can write

$$\frac{Q}{C} = \mathscr{E}(t)$$

This is a relationship between the (magnitude of the) charge on each plate. To determine how the current changes with time we differentiate both sides of the above result with respect to time and use the definition of current (3.1) (see also the comments in Section 11.4). We find the desired result is given as

$$I = C \times \frac{d\,\mathscr{E}(t)}{dt}$$

Note that this equation relates the current to the *rate of change* of the e.m.f. with respect to time, and not to the e.m.f. directly. This has very important implications, as we now discuss. For the case of $\mathcal{E}(t)$ the sinusoidal wave form (12.2) we have

$$I(t) = C \times \frac{d\,\mathcal{E}(t)}{dt} = C \times \frac{d}{dt}\,\mathcal{E}_0 \sin(2\pi ft) = \mathcal{E}_0 \times (2\pi fC)\cos(2\pi ft)$$

Making use of well known properties of the sine and cosine functions, we can write the above equation in the form

$$I(t) = I_0 \sin\left(2\pi ft + \frac{\pi}{2}\right)$$

where we have also introduced the maximum or peak current through the circuit, $I_0$, which is related to $\mathcal{E}_0$, $C$ and $f$ as

$$I_0 = \mathcal{E}_0 \times (2\pi fC)$$

There are two important points to note here:

*(1)* The current varies with the same frequency as the source of the e.m.f., but it is no longer in phase with the e.m.f. Here the current *leads* the e.m.f. by $\pi/2$. Alternatively, we can argue that the e.m.f. *lags* the current by $\pi/2$. This is understandable since, if a capacitor is uncharged, the p.d. across it is zero. Therefore, for a p.d. to develop across the capacitor, a current must first flow to charge the capacitor. This is shown in Figure 12.16.

*(2)* If we write the relationship between the peak current through the circuit, $I_0$, and the peak p.d. across the capacitor, $\mathcal{E}_0$, in the following form:

$$I_0 = \frac{\mathcal{E}_0}{X_C}, \quad \text{where } X_C = \frac{1}{2\pi fC} \qquad (12.9)$$

which is again very suggestive of an Ohm's law type relationship. The quantity $X_C$ seems to be a resistance-like property of the circuit, and is referred to as the capacitive reactance. Note that the reactance depends inversely on frequency; see Figure 12.17.

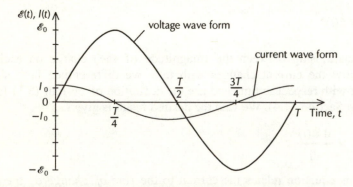

*Figure 12.16*   **The voltage and current wave forms for a capacitor**

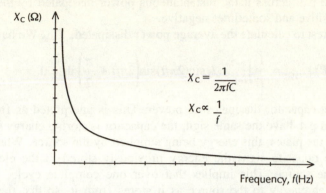

*Figure 12.17* **The capacitive reactance of a capacitor versus the frequency**

Note that as the frequency gets smaller the reactance becomes larger. In the limit of a zero frequency source, the reactance is infinite and the capacitor acts like an open circuit. At the opposite extreme, as the frequency of the supply increases the reactance gets smaller – the capacitor is loosely said to offer a smaller resistance to the flow of current. In the limit of arbitrarily large frequency the capacitor acts like a short circuit.

### Power in a pure capacitive AC circuit

The instantaneous power developed across the capacitor is given by (6.5). We have

$$P(t) = I(t) \times V_C(t)$$

Figure 12.18 shows how the instantaneous power dissipated by the capacitor varies with time over one complete cycle. Since the current through the capacitor is out of

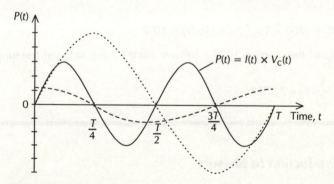

*Figure 12.18* **The power developed across a capacitor over one complete cycle. Note that the power is no longer always positive**

phase with the p.d. across it the instantaneous power dissipated by the capacitor is sometimes positive and sometimes negative.

It is of interest to calculate the average power dissipated, $P_{ave}$. We have

$$P_{ave} = \langle P \rangle_{\text{one cycle}} = \frac{1}{T} \int_{t=0}^{t=T} \varepsilon_0 I_0 \sin(2\pi ft) \sin\left(2\pi ft + \frac{\pi}{2}\right) dt = 0$$

Thus a perfect capacitor dissipates no power. This is interpreted as follows: When the current and p.d. have the same sign, the capacitor is storing energy in the electric field between the plates, this energy being supplied by the source. When the current and p.d. have opposite signs the energy previously stored in the electric field is returned to the source. This implies that over one complete cycle, the capacitor returns as much energy to the source as it stores from it, so that the *total* power dissipated by the capacitor over one cycle is zero.

━━━━━ **WORKED EXAMPLE 12.4**
_____

A 5 μF capacitor is placed in series with a sinusoidal supply:

$$\mathscr{E}(t) = \mathscr{E}_0 \sin(2\pi ft)$$

At what frequency is the capacitive reactance, $X_C$, equal to 50 Ω? If the peak current through the circuit at this frequency is 200 mA, determine the r.m.s. e.m.f. of the supply.

− The capacitive reactance is given as

$$X_C = \frac{1}{2\pi fC}$$

from which we may determine the frequency as

$$f = \frac{1}{2\pi X_C C} = \frac{1}{2 \times 3.142 \times (50\ \Omega) \times (5 \times 10^{-6}\ \text{F})} = 636.620\ \text{Hz}$$

Since the peak current through the circuit is 200 mA, it follows that the peak e.m.f. of the supply is

$$\mathscr{E}_0 = I_0 X_C = (200 \times 10^{-3}\ \text{A}) \times (50\ \Omega) = 10\ \text{V}$$

The peak voltage of the supply is 10 V. It follows that the r.m.s. value of the supply is given as

$$\mathscr{E}_{\text{r.m.s.}} = \frac{1}{\sqrt{2}} \mathscr{E}_0 = 7.071\ \text{V}$$

━━━━━━━━━━━━━━━━━━━━━━━━━━━━━━━━━━━━━━━━━━━━━━━━

## 12.5   An introduction to phasors

Alternating currents and voltages are conveniently represented by rotating vectors called phasors. The phasor rotates anti-clockwise with the same frequency, $f$, as the wave form it represents, and the length of the phasor is equal to the amplitude or peak value of the wave form.

Figure 12.19 depicts the phasor which represents a sinusoidal wave form of amplitude $A$ and frequency $f$:

$$y(t) = A \sin(2\pi ft)$$

Note that at time $t = 0$ the phasor is horizontal. This direction conveniently forms a reference direction from which to measure the phase or the phase angle, where the phase angle is given by the value $2\pi ft$. As time increases the phasor rotates in the anti-clockwise direction and the angle the phasor makes with respect to the horizontal is given by the phase of the wave form.

Let us now consider the following two sinusoidal wave forms:

$$y_1(t) = A_1 \sin(2\pi ft) \tag{12.10a}$$
$$y_2(t) = A_2 \sin(2\pi ft + \phi) \tag{12.10b}$$

These two wave forms have the same frequency but have different amplitudes. Also, the wave form $y_2(t)$ leads the wave form $y_1(t)$ by the phase shift $\phi$.

Let us consider the phasor diagrams for each wave form, shown in Figure 12.20. Notice that the respective phasors both rotate with the same frequency so that the angle between them is constant and given by the phase shift between the wave forms, $\phi$. Therefore, if we are just interested in the phase relations between $y_1$ and $y_2$ we can simply ignore the fact that the two phasors are rotating and treat the phasor simply like familiar vectors. This is shown in Figure 12.21. The time dependence of the phasors can be brought back simply by rotating these vectors in the anti-clockwise direction with the appropriate frequency. This will be useful for us when we discuss the phase relationships between the current through a component and the p.d. across it.

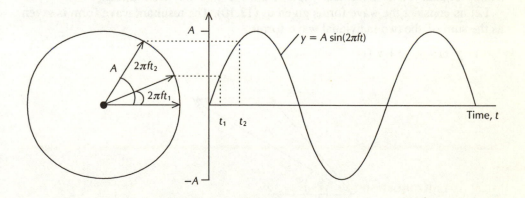

*Figure 12.19*   **The vector of length *A* rotating with angular frequency ω = 2πf represents the sinusoidal function *y*(*t*) = *A* sin(2πft)**

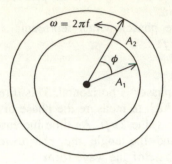

*Figure 12.20* **The phasor diagrams for the two wave forms $y_1(t) = A_1 \sin(2\pi ft)$ and $y_2(t) = A_2 \sin(2\pi ft + \phi)$**

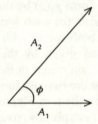

*Figure 12.21* **Ignoring the time dependence of phasors and treating them like vectors**

Another important use of phasor diagrams is to determine the effect of adding two wave forms. Again we can use the fact that the two phasors are rotating with the same frequency, and conveniently ignore the fact that they are rotating.

Let us consider the wave forms given in (12.10). The resultant wave form is given as the sum of the two individual wave forms

$$y(t) = y_1(t) + y_2(t)$$

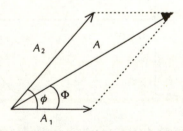

*Figure 12.22* **The resultant of two phasors**

We write this wave form as a sinusoidal wave form of the same frequency, $f$, as the original wave forms, but of amplitude $A$ and with a phase shift $\phi$

$$y(t) = A \sin(2\pi ft + \phi)$$

The quantities $A$ and $\phi$ can be most easily determined from the phasor diagram as the vector addition of the phasors representing the wave form $y_1$ and $y_2$, see Figure 12.22.

Worked Example 12.5 illustrates some of these concepts.

━━━ **WORKED EXAMPLE 12.5**

Determine the resultant of the two wave forms

$$y_1(t) = 5 \sin(200\pi t)$$

$$y_2(t) = 8 \sin\left(200\pi t + \frac{\pi}{3}\right)$$

— We note that the two wave forms have the same frequency, $f = 100$ Hz. With reference to Figure 12.22, we have that

$$A \sin(\Phi) = A_2 \sin(\phi) = 8 \sin\left(\frac{\pi}{3}\right) = 6.928$$

$$A \cos(\Phi) = A_1 + A_2 \cos(\phi) = 5 + 8 \cos\left(\frac{\pi}{3}\right) = 9$$

If we divide these two results we find

$$\frac{A \sin(\Phi)}{A \cos(\Phi)} = \tan(\Phi) = \frac{6.928}{9} = 0.770$$

from which the phase shift $\Phi$ is given as

$$\Phi = \tan^{-1}(0.770) = 0.656 \text{ rad}\,(= 37.5°)$$

If we square and add the two results we find

$$(A \sin(\Phi))^2 + (A \cos(\Phi))^2 = A^2 = 5^2 + 6.928^2 = 72.997$$

from which the peak value or the amplitude of the wave form is given as

$$A = 8.544$$

Therefore the resultant wave form is given as

$$y(t) = 8.544 \sin(200\pi t + 0.656)$$

Note that phase shifts are usually quoted in radians rather than degrees. The relationship between the radian and the degree is given in the Appendix.

Table 12.1

| Circuit element | Peak current, p.d. | Phase relation |
|---|---|---|
| Resistor | $I_0 = \mathcal{E}_0/R$ | $\mathcal{E}$, $I$ in phase |
| Inductor | $I_0 = \mathcal{E}_0/X_L$, $X_L = 2\pi fL$ | $I$ lags $\mathcal{E}$ by $\pi/2$ |
| Capacitor | $I_0 = \mathcal{E}_0/X_C$, $X_C = 1/2\pi fC$ | $I$ leads $\mathcal{E}$ by $\pi/2$ |

## 12.6 Relationship with phasor diagrams

Table 12.1 summarizes the relationship between the peak current through and the p.d. across the components of interest, and the phase relationship between the current and the p.d.

A convenient mnemonic for recalling the phase shifts between current and voltage in capacitive and inductive circuits is

### *CIVIL*

In a capacitive (*C*) circuit *I* before *V* (i.e. *I* leads *V*), *V* before *I* in an inductive (*L*) circuit (i.e. *V* leads *I*).

It is important to note that while inductive and capacitive reactances are clearly resistance-like properties, they are not the same as resistance. The two important differences are that

- pure inductive and capacitive circuits do not dissipate power, and
- the current through an inductive or a capacitive circuit is out of phase with the supply.

The wave forms and phasor diagrams for current through and p.d. across the resistor, inductor and capacitor are shown in Figures 12.23(a), (b) and (c), respectively. Phasor diagrams will be found to be very useful when discussing AC circuits consisting of two or more components.

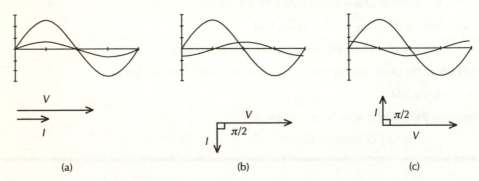

*Figure 12.23*  **The voltage and current wave forms and phasor diagrams for (a) a resistor, (b) an inductor and (c) a capacitor**

## 12.7 Summary

We began this chapter with a discussion of the sinusoidal wave form, introducing the concept of average value and phase shift.

We then discussed the behaviour of circuits consisting of a resistor, inductor and capacitor separately connected to an alternating voltage supply. Our interest lay in determining the frequency response of such circuits connected to a sinusoidal supply. The frequency response of a circuit simply gives us information about the relationship between the current through the circuit and the p.d. across it. For the inductor and capacitor we found that we were able to introduce a resistance-like property which limits the current in the circuit in a manner similar to that in which the resistance limits the current in a circuit. This property was referred to as the inductive and capacitive reactances, respectively. Whereas resistance is a property of a resistor that is independent of the frequency of the supply, we found that both the inductive and capacitive reactances are properties that are frequency dependent. In particular, the inductive reactance is proportional to the frequency, and the capacitive reactance is inversely proportional to the frequency of the supply.

The phasor diagram was introduced as a geometric means of representing a sinusoidal wave form. Such diagrams were found to be useful when we

- determine the effect of adding two wave forms, and
- discuss the phase shift between a current through a component and the p.d. across it.

From our work we were able to determine the relevant phasor diagrams for the resistor, inductor and capacitor. These features, in particular, will be found to be very useful when discussing AC circuits consisting of two or more components.

## ▬ Problems

**12.1**  An AC generator consists of a coil consisting of 100 turns of wire rotating with an angular speed of $\omega = 314$ rad s$^{-1}$ in a uniform magnetic field of $B = 1.5$ T. If the area of the coil is 200 mm$^2$ determine the peak, the peak-to-peak and the r.m.s. value of the e.m.f. induced and the frequency, $f$, of the supply. What is the period, $T$, of the supply?

*Hint:* Make use of (9.7) and note that we are given the area of the coil. Also take into account that the coil consists of more than just one turn as assumed in the derivation of (9.7).

o **12.2**  Determine the frequency, period and the average value for each of the following functions of time:

(a) $\mathcal{F}(t) = 100 + 3 \sin(2\pi t/5)$ $0 \leqslant t \leqslant 5$ s, (b) $\mathcal{F}(t) = 2 \sin(10\pi t)$, $0.1 \leqslant t \leqslant 0.3$ s.

o **12.3**  Verify the result (12.6a) in the text.

*Hint:* Make use of the results of the Appendix.

**12.4** Use a phasor diagram to determine the resultant when the following two wave forms are added:

$$\mathscr{E}_1(t) = 5\sin(200\pi t)$$

$$\mathscr{E}_2(t) = 10\sin\left(200\pi t + \frac{\pi}{3}\right)$$

**12.5** A 5 µF capacitor is connected to a 50 Hz supply of peak value 100 V. Determine the capacitive reactance and the peak current that flows. Repeat if the capacitor is replaced by a 200 mH inductor.

# CHAPTER THIRTEEN

# *Series AC circuits*

In Chapter 12 we made use of Kirchhoff's laws to determine the frequency response of a resistor, an inductor and a capacitor when connected to a sinusoidal voltage supply.

We found that the current through a resistor was in phase with the p.d. across it, i.e. as the voltage changed across the resistor the current changed exactly in the same manner. This implies that resistance is a property of a resistor which does not vary with the frequency of the driving e.m.f. This contrasts with the behaviour of an inductor or a capacitor, where we found the current and voltage were no longer in phase with each other. In particular, we found that the current lags the voltage by a phase angle $\pi/2$ for an inductor and leads the voltage by $\pi/2$ for a capacitor. We also noted a resistance-like property of inductors and capacitors which depends upon the frequency of the e.m.f; this is termed the reactance.

Following this we introduced the concept of the phasor and the phasor diagram. Phasors are a geometrical way of representing time-varying quantities, such as sine waves, in terms of their magnitude and phase. We found it was possible to interpret previous results in terms of phasors and phasor diagrams.

We now wish to investigate the frequency response, i.e. the phase relationship between current and voltage and the resistance-like property, for the following circuits: a resistor and inductor in series (RL circuit), a resistor and capacitor in series (RC circuit) and an inductor and a capacitor in series (LC circuit). We will find a very interesting property of the LC circuit, referred to as resonance, which we take up in some detail when we generalize to the RLC circuit. In the following we will make extensive use of the phasor diagram to analyze each circuit.

## 13.1  The series RL circuit

Consider the circuit shown in Figure 13.1, consisting of a resistance $R$ and an inductance $L$ connected in series to a sinusoidal e.m.f.

**233**

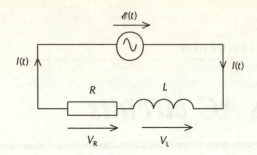

*Figure 13.1*    **The effects of a sinusoidal supply on a series resistive and inductive circuit**

## The phasor diagram approach

Figures 13.2(a) and (b) show the phasor diagrams for $R$ and $L$ separately. To combine the two phasor diagrams it is first necessary to rotate the phasor diagram representing the current through and the p.d. across the inductor such that this current vector is in the same direction as the current vector for the resistor, given in Figure 13.2(a). The result of this rotation is shown in Figure 13.2(c).

The supply voltage, $\mathcal{E}$, is, at all times, the instantaneous sums of the p.d. across $R$ and the p.d. across $L$ – this is just Kirchhoff's voltage law. However, since these voltages are $\pi/2$ out of phase, we must use their vector sum to combine the peak values. From Figure 13.2(d) we have

$$\mathcal{E}_0 = \sqrt{V_R^2 + V_L^2}$$

Recall (Figure 13.2) that the length of a phasor is the peak value of the wave form it represents, so that $V_R$ and $V_L$ are the peak values of the p.d.s across the resistor and the inductor, respectively.

We can also deduce from the figure that the current lags the supply by the phase shift $\phi$ determined from

$$\tan(\phi) = \frac{V_L}{V_R}$$

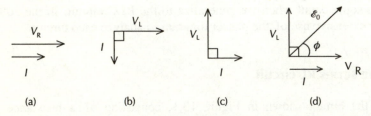

|     |     |     |     |
| (a) | (b) | (c) | (d) |

*Figure 13.2*    **The phasor diagrams for $R$ and $L$ separately, and their vector combination**

Let us consider Figure 13.2(d) in some detail. Suppose we divide the length of each phasor by $I_0$, the peak current through the circuit. Again, recall that $V_L$ and $V_R$ are the peak values of the p.d.s across the inductor and the resistor, respectively, so that dividing these values by the peak current through each component gives the resistance and the inductive reactance, respectively. We have that

$$R = \frac{V_R}{I_0} \quad \text{and} \quad X_L = \frac{V_L}{I_0}$$

This leads us to the concept of the *impedance diagram*, shown in Figure 13.3. In the figure we have defined the *impedance, Z*, by

$$Z = \frac{\mathcal{E}_0}{I_0}$$

The impedance is clearly a resistance-like property of the circuit since it determines the peak value of the current through the circuit for a given source of e.m.f.

From the impedance diagram we have that the impedance of the circuit is related to the resistance, $R$, and the inductive reactance, $X_L$, by

$$Z = \sqrt{R^2 + X_L^2}, \quad X_L = 2\pi f L \tag{13.1}$$

We also have that the phase shift $\phi$ is determined as the ratio of the reactance to the resistance,

$$\tan(\phi) = \frac{X_L}{R} \tag{13.2}$$

Notice that in the limit of small reactance (either the frequency, $f$, of the supply is small or the inductance, $L$, is small) the impedance is equal to the resistance of the circuit and the phase shift is small. This is reasonable since, in this limit, the inductance has little effect, and the circuit behaves similar to a pure resistive circuit. That the current is in phase with the supply is well known for such a circuit. At the other extreme, in the limit of small resistance, the circuit is similar to a pure inductive circuit, the impedance is just given by the inductive reactance and the phase shift approaches $\pi/2$, which is expected.

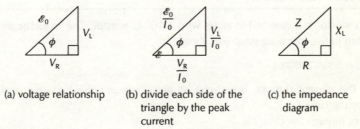

(a) voltage relationship

(b) divide each side of the triangle by the peak current

(c) the impedance diagram

*Figure 13.3*   **The stages leading to the impedance diagram**

### The power dissipated in an RL circuit

The instantaneous power developed across the circuit is given as the product of the current through and the p.d. across the circuit. We find that the power dissipated across the circuit is given as

$$P_{ave} = \langle P \rangle_{one\ cycle} = \frac{1}{T} \int_{t=0}^{t=T} \mathscr{E}_0 I_0 \sin(2\pi ft)\sin(2\pi ft - \phi)\ dt$$

Evaluating this yields the result

$$P_{ave} = \frac{1}{2} \mathscr{E}_0 I_0 \cos(\phi) = \mathscr{E}_{r.m.s.} I_{r.m.s.} \cos(\phi)$$

Note that in this circuit power is dissipated. Making use of the relationship between the phase angle $\phi$ and $R$ *and* $X_L$, we can write

$$P_{ave} = \mathscr{E}_{r.m.s.} I_{r.m.s.} \frac{R}{Z}$$

which clearly identifies the resistor as the reason for the non-zero result. This is not too surprising, since we found that there is no power dissipated in a pure inductive circuit.

The factor $\cos(\phi)$ is referred to as the *power factor* and is usually defined as

$$power\ factor = \frac{true\ power}{apparent\ power}$$

where the true power dissipated is the average power $P_{ave}$, while the apparent power dissipated by the circuit is given as the product of $\mathscr{E}_{r.m.s.}$ and $I_{r.m.s.}$.

The apparent power is the power, we might argue, the circuit is dissipating, given that the current through the circuit and the p.d. across it are $I_{r.m.s.}$ and $\mathscr{E}_{r.m.s.}$, respectively. The reason for this difference is that the current and the p.d. are not in phase. Thus the product of the r.m.s. current and voltage gives the maximum power dissipation, but the power available is mediated by the factor $\cos(\phi)$.

━━━ **WORKED EXAMPLE 13.1**

A 200 mH inductor is connected in series with a 100 Ω resistor. The inductor and the resistor are connected to the following sinusoidal supply

$$\mathscr{E}(t) = 5 \sin(1000\pi t)$$

Determine the current through the circuit at any time and the apparent and true power dissipated by the circuit.

– The current through the circuit is

$$I(t) = I_0 \sin(2\pi ft - \phi)$$

We need to determine the peak current, $I_0$, and the phase shift, $\phi$. The inductive reactance is

$$X_L = 2\pi fL = 2 \times 3.142 \times (500\ \text{Hz}) \times (200 \times 10^{-3}\ \text{H}) = 628.319\ \Omega$$

since the frequency of the supply is 500 Hz. It follows that the impedance of the circuit is

$$Z = \sqrt{R^2 + X_L^2} = 636.227\Omega$$

Given the impedance the peak current through the circuit is

$$I_0 = \frac{\mathscr{E}_0}{Z} = \frac{5\ \text{V}}{636.227\Omega} = 7.859\ \text{mA}$$

The phase shift $\phi$ is determined from

$$\tan(\phi) = \frac{X_L}{R} = \frac{628.319}{100} = 6.283\ 19$$

giving

$$\phi = \tan^{-1}(6.283\ 19) = 1.413\ \text{rad}\ (=81°)$$

Thus, the current through the circuit at any instant is

$$I(t) = 7.859\ \sin(1000\pi t - 1.413)\ \text{mA}$$

The apparent power dissipated by the circuit is given as the product of the r.m.s. e.m.f. and current and is

$$P_{\text{apparent}} = \mathscr{E}_{\text{r.m.s.}} I_{\text{r.m.s.}} = \tfrac{1}{2}\mathscr{E}_0 I_0 = \tfrac{1}{2}\,(5\ \text{V}) \times (7.859 \times 10^{-3}\ \text{A}) = 19.648\ \text{mW}$$

Because the current through the circuit is out of phase with the supply, the true power dissipated by the circuit is

$$P_{\text{ave}} = P_{\text{apparent}} \cos(\phi) = 3.088\ \text{mW}$$

since the power factor is

$$\text{power factor} = \cos(\phi) = 0.157$$

---

## 13.2 The series RC circuit

Figure 13.4 shows a circuit consisting of a resistance $R$ and a capacitance $C$ connected in series to a sinusoidal e.m.f.

### The phasor diagram approach

The arguments presented for the RL case may be given again, except we replace 'L' by 'C'. Thus, Figure 13.5 shows the phasor diagrams for $R$ and $C$ and their

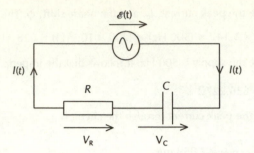

**Figure 13.4**   **The effects of a sinusoidal supply on a series resistive and capacitive circuit**

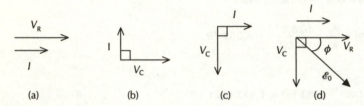

**Figure 13.5**   **The phasor diagrams for *R* and *C* separately, and their vector combination**

combination. Again, note that to combine the two phasor diagrams it is first necessary to rotate the phasor diagram representing the current through and p.d. across the capacitor, such that the current vector is in the same direction as the current vector for the resistor, given in Figure 13.5(a). The result of this rotation is shown in Figure 13.5(c).

From Figure 13.5(d) the relationship between the peak values of the source and the p.d.s across the resistor and capacitor are given as

$$\mathscr{E}_0 = \sqrt{V_R^2 + V_C^2}$$

We can also deduce from the figure that the current leads the supply by the phase shift $\phi$ determined from

$$\tan(\phi) = \frac{V_C}{V_R}$$

Similar considerations to those given previously lead to the impedance diagram for the RC circuit as shown in Figure 13.6. The impedance $Z$ is clearly related to the resistance, $R$, and the capacitive reactance, $X_C$, by the result

$$Z = \sqrt{R^2 + X_C^2}, \qquad X_C = \frac{1}{2\pi f C} \tag{13.3}$$

and, again, is clearly a resistance-like property of the circuit.

The phase shift $\phi$ is determined from

$$\tan(\phi) = \frac{X_C}{R} \tag{13.4}$$

*Figure 13.6* **The impedance diagram**

Notice that in the limit of small reactance (either the frequency, $f$, of the supply is large, or the capacitance, $C$, is large) the impedance is equal to the resistance of the circuit and the phase angle is small. This is reasonable since in this limit the capacitance has little effect and the circuit looks similar to a pure resistive circuit. For such a circuit it is well known that the current is in phase with the supply. At the other extreme, in the limit of small resistance, the circuit is a pure capacitive circuit

**▬▬ WORKED EXAMPLE 13.2**

A 100 $\Omega$ resistor is connected in series to a 25 µF capacitor to the following sinusoidal supply:

$$\mathscr{E}(t) = 25 \sin(500\pi t) \text{ V}$$

Determine the current through the circuit at any time.
— The current through the circuit is

$$I(t) = I_0 \sin(2\pi f t + \phi)$$

We need to determine the peak current, $I_0$, and the phase shift, $\phi$. The capacitive reactance is

$$X_C = \frac{1}{2\pi f C} = \frac{1}{2 \times 3.142 \times (250 \text{ Hz}) \times (25 \times 10^{-6} \text{ F})} = 25.465 \ \Omega$$

since the frequency of the supply is 250 Hz. It follows that the impedance of the circuit is

$$Z = \sqrt{R^2 + X_C^2} = 103.191 \ \Omega$$

Given the impedance the peak current through the circuit is

$$I_0 = \frac{\mathscr{E}_0}{Z} = \frac{25 \text{ V}}{103.191 \Omega} = 0.242 \text{ A}$$

The phase shift $\phi$ is determined from

$$\tan(\phi) = \frac{X_L}{R} = \frac{25.465}{100} = 0.254\,65$$

giving

$$\phi = \tan^{-1}(0.254\,65) = 0.249 \text{ rad } (=14°)$$

Thus, the current through the circuit at any instant is

$$I(t) = 0.242 \sin(500\pi t + 0.249) \text{ A}$$

the impedance is equal to the reactance and the phase angle approaches $\pi/2$, which is expected.

### The power dissipated in an RC circuit

In a similar approach to that discussed for the RL circuit, it can be shown that the average power dissipated in an RC circuit is given as

$$P_{ave} = \mathcal{E}_{r.m.s.}I_{r.m.s.} \cos(\phi)$$

which can, again, be written in terms of the resistance and the impedance as

$$P_{ave} = \mathcal{E}_{r.m.s.}I_{r.m.s.} \frac{R}{Z}$$

again, clearly identifying the resistor as the origin of the non-zero result. Similarly, it can be argued that this is not too surprising, since we found that there is no power dissipated in a pure capacitive circuit.

## 13.3   The series LC circuit

Consider the circuit shown in Figure 13.7 consisting of an inductance $L$ and a capacitance $C$ connected in series to a sinusoidal e.m.f.

Figures 13.8(a) and (b) show the phasor diagrams for the current and p.d. across $L$ and $C$ separately. As usual, to combine them we must rotate one of the phasor diagrams such that the current vector for the inductor and capacitor are in the same direction. This is shown in Figure 13.8(c), from which we note that the p.d. across the inductor is exactly out of phase with the p.d. across the capacitor.

Note that the supply voltage, $\mathcal{E}_0$, is simply given as

$$\mathcal{E}_0 = \begin{cases} V_L - V_C & \text{if } V_L > V_C \\ V_C - V_L & \text{if } V_C > V_L \end{cases}$$

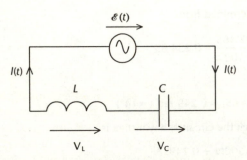

*Figure 13.7*   **The effects of a sinusoidal supply on a series inductive and capacitive circuit**

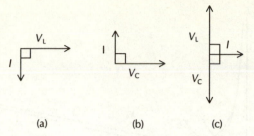

(a)                    (b)                    (c)

*Figure 13.8*   **The phasor diagrams for *L* and *C* separately, and their vector combination**

and the phase shift $\phi$ is given as

$$\phi = \begin{cases} +\dfrac{\pi}{2} & \text{if } V_L > V_C \\[2mm] -\dfrac{\pi}{2} & \text{if } V_C > V_L \end{cases}$$

i.e. either the supply voltage leads the current by $\pi/2$ if $V_L > V_C$, or it lags the current by $\pi/2$ if $V_C > V_L$. These conditions can be stated in terms of the inductive and capacitive reactances as

$$\phi = \begin{cases} +\dfrac{\pi}{2} & \text{if } X_L > X_C \\[2mm] -\dfrac{\pi}{2} & \text{if } X_C > X_L \end{cases}$$

The impedance diagram follows from Figure 13.8(c). We find that the impedance of the circuit is wholly reactive, this is not surprising since there are no resistive elements, and given as

$$Z = \begin{cases} X_L - X_C & \text{if } X_L > X_C \\ X_C - X_L & \text{if } X_C > X_L \end{cases}$$

### Resonance

Recall that the inductive reactance is directly proportional to the frequency (see Equation (12.8)), while the capacitive reactance is inversely proportional to the frequency (see Equation (12.9)). It thus follows that $X_L$ increases as the frequency of the supply increases, while $X_C$ decreases (recall Figures 12.14 and 12.17, respectively).

Figure 13.9 shows how the impedance of the LC circuit varies with the frequency (also shown are the inductive and capacitive reactances). From the figure we note

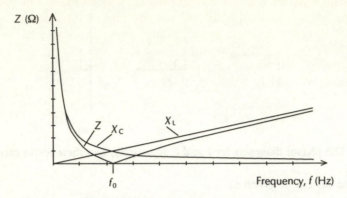

*Figure 13.9*   **The variation of impedance for an LC circuit with frequency**

that the impedance has a minimum value of zero at the frequency $f_0$. At this frequency the LC circuit offers no restriction to the current flow. This phenomenon is known as *resonance*.

Let us determine the *resonant frequency*, $f_0$. At resonance the inductive reactance is equal to the capacitive reactance:

$$2\pi f_0 L = \frac{1}{2\pi f_0 C}$$

This equation can easily be solved for the resonant frequency. We find

$$f_0 = \frac{1}{2\pi\sqrt{LC}} \tag{13.5}$$

## The power dissipated in an LC circuit

Using an approach similar to that discussed for the RL and RC circuits it can be shown that the average power dissipated in an LC circuit is zero. This is not too surprising since we have previously emphasized that power dissipation can be traced to the resistive elements in the circuit, which are absent here.

## 13.4   The series RLC circuit – resonance revisited

For the series RL circuit it was noted that at a particular frequency, the resonant frequency, the impedance of the circuit has a minimum value of zero, i.e. at this frequency the LC circuit offers no restriction to the current flow. This implies that the current through the circuit would rise indefinitely. In practice there will always be some resistive elements in the circuit present, which would limit the current. In the following we discuss further the LC circuit, introducing a resistive component. Figure 13.10 shows a resistor, inductor and capacitor connected in series to a sinusoidal e.m.f.

An analysis of the circuit by means of the phasor diagram approach is straightforward. We find the following results:

- The impedance of the circuit, $Z$, is

$$Z = \sqrt{R^2 + (X_L - X_C)^2} \qquad (13.6)$$

The variation of the impedance with the frequency is depicted in Figure 13.11.
- The current lags the voltage supply by the phase shift $\phi$, where the phase shift is determined from

$$\tan(\phi) = \frac{X_L - X_C}{R}$$

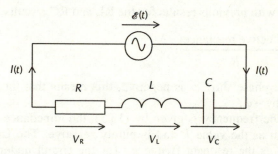

Figure 13.10   **The effects of a sinusoidal supply on a series resistive, inductive and capacitive circuit**

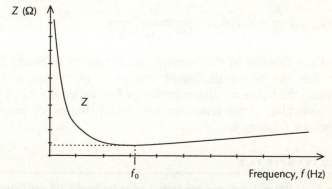

Figure 13.11   **The variation of impedance for an RLC circuit with frequency. Note that the resonant frequency has not changed, but that the variation of impedance with frequency is smoother around the resonant frequency**

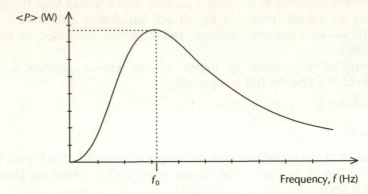

*Figure 13.12* **The variation of the average power dissipated by an RLC circuit with frequency. Note that the maximum power is dissipated at the resonant frequency**

Comparing this with previous results for the RL and RC circuits, we can write

$$\tan(\phi) = \frac{\text{effective reactance}}{R} \tag{13.7}$$

Note that if the phase shift $\phi$ is negative, this means that the current *leads* the voltage supply by $-\phi$.

Note that at the frequency $f_0$ given by (13.5), the impedance of the circuit is a minimum, given as the value $R$ and is purely resistive. The frequency (13.5) is still referred to as the resonant frequency for the circuit under discussion. Note that the resonant frequency only depends upon $L$ and $C$, and is independent of the circuit resistance $R$.

■ The average power dissipated by the circuit over one cycle, $P_{ave}$, is given as

$$P_{ave} = \mathscr{E}_{r.m.s.} I_{r.m.s.} \frac{R}{Z} = \mathscr{E}_{r.m.s.} I_{r.m.s.} \frac{R}{\sqrt{R^2 + (X_L - X_C)^2}} \tag{13.8}$$

and is plotted as a function of the frequency of the source in Figure 13.12.

We note that the power dissipated reaches a maximum at the resonant frequency. From the figure we also note that as $R$ is varied the width of the curve changes. In particular, as the resistance gets larger, the curve becomes broader about the resonant frequency.

━━━ **WORKED EXAMPLE 13.3**

A circuit consists of a 200 mH inductor connected in series to a 5 µF capacitor and a 100 Ω resistor. The circuit is then connected to a 250 Hz sinusoidal supply of peak e.m.f. 10 V, i.e.

$$\mathscr{E}(t) = 10 \sin(500\pi t) \text{ V}$$

Determine the current through the circuit at any time and the apparent and true power dissipated by the circuit. Determine the resonant frequency of the circuit, the inductive and capacitive reactances and the impedance of the circuit. What current flows through the circuit if connected to a sinusoidal supply of this frequency and of peak e.m.f. 10 V?
– The current through the circuit is

$$I(t) = I_0 \sin(2\pi f t - \phi)$$

and we need to determine the peak current, $I_0$, and the phase shift, $\phi$.

We first determine the inductive and capacitive reactances and the result for the impedance follows. The frequency of the supply is 250 Hz, so that

$$X_L = 2\pi f L = 314.159 \ \Omega \quad \text{and} \quad X_C = \frac{1}{2\pi f C} = 127.324 \ \Omega$$

It follows that the impedance of the circuit is

$$Z = \sqrt{R^2 + (X_L - X_C)^2} = 211.914 \ \Omega$$

Given the impedance, the peak current is given as

$$I_0 = \frac{\mathscr{E}_0}{Z} = \frac{10 \ \text{V}}{211.914 \Omega} = 47.189 \ \text{mA}$$

The phase shift $\phi$ is determined from

$$\tan(\phi) = \frac{X_L - X_C}{R} = \frac{186.835}{100} = 1.868 \ 35$$

giving

$$\phi = \tan^{-1}(1.868 \ 35) = 1.079 \ \text{rad} \ (= 62°)$$

Thus, the current through the circuit at any instant is

$$I(t) = 47.189 \sin(500\pi t - 1.079) \ \text{mA}$$

The apparent power dissipated by the circuit is given as the product of the r.m.s. e.m.f. and current and is

$$P_{\text{apparent}} = \mathscr{E}_{\text{r.m.s.}} I_{\text{r.m.s.}} = \tfrac{1}{2}\mathscr{E}_0 I_0 = \tfrac{1}{2}(10 \ \text{V}) \times (47.189 \times 10^{-3} \ \text{A}) = 0.236 \ \text{W}$$

Because the current through the circuit is out of phase with the supply, the true power dissipated by the circuit is

$$P_{\text{ave}} = P_{\text{apparent}} \cos(\phi) = 0.111 \ \text{W}$$

since the power factor is

$$\text{power factor} = \cos(\phi) = 0.472$$

The resonant frequency, $f_0$, is

$$f_0 = \frac{1}{2\pi\sqrt{LC}} = \frac{1}{2\pi\sqrt{(200 \times 10^{-3} \ \text{H}) \times (5 \times 10^{-6} \ \text{F})}} = 159.155 \ \text{Hz}$$

At this frequency the inductive reactance is

$$X_L = 2\pi f L = 200 \ \Omega$$

and, from the definition of the resonant frequency, we have

$$X_C = 200 \ \Omega$$

The impedance of the circuit is just given by the resistance $R$,

$$Z = R = 100 \ \Omega$$

since the effects of the inductive and capacitive reactances cancel each other out.

The current through the circuit at resonance is

$$I(t) = 100 \sin(318.31\pi t) \ \text{mA}$$

since the impedance is purely resistive and therefore the current is in phase with the supply.

## 13.5  Summary

We discussed the frequency responses of circuits connected to a sinusoidal supply. The frequency response simply gives us information about the relationship between the current through the circuit and the p.d. across the circuit. We discussed RL, RC and LC circuits. In particular, the resonant frequencies of the series LC and RLC circuits were discussed.

## ▬▬ Problems

**13.1**  A circuit consists of a 200 $\Omega$ resistor and an inductor connected in series to the following supply:

$$\mathscr{E}(t) = 25 \sin(100\pi t) \ \text{V}$$

If the peak current through a circuit is 20 mA, determine the impedance of the circuit, the inductive reactance and the inductance $L$. Does the current lead or lag the voltage? Determine the phase shift.

**13.2**  A circuit consists of a 1 k$\Omega$ resistor and a 10 µF capacitor connected in series to the following supply:

$$\mathscr{E}(t) = 50 \sin(500\pi t) \ \text{V}$$

Determine the capacitive reactance, the impedance and the peak current through the circuit. Does the current lead or lag the voltage? Determine the phase shift.

o **13.3**  A 5 mH inductor is connected in series to a 10 µF capacitor. Determine the range of frequencies for which the phase shift is $\pi/2$ and those frequencies for which the phase shift is $-\pi/2$.

o **13.4** Show that $\sqrt{(LC)}$ has the units of time.
*Hint:* Make use of the results in Chapter 1.

**13.5** A circuit consists of a 250 Ω resistor, a 100 mH inductor and a 20 μF capacitor connected in series to a sinusoidal voltage supply. If the frequency of the supply is the resonant frequency of the circuit, determine the impedance of the circuit.

**13.6** A 100 Ω resistor, a 10 mH inductor and a 2 μF capacitor are connected in series to the following supply:

$\mathscr{E}(t) = 10 \sin(2000\pi t)$ V

(a) Determine the inductive and capacitive reactances and the impedance of the circuit.

(b) Determine the current through the circuit.

(c) Determine the power factor and the power dissipated by the circuit.

(d) Determine the resonant frequency for the circuit.

(e) Complete parts (a), (b) and (c) if the frequency of the source is changed to the resonant frequency, but the peak value of the voltage source remains unchanged.

# CHAPTER FOURTEEN

# *Semiconductors and the PN junction*

In this chapter we introduce semiconductors, beginning with a brief discussion of charge movement in an *intrinsic* semiconductor. We will find that in an intrinsic semiconductor there are two types of charge movement which contribute to an electric current: the familiar electron flow we have previously encountered with metallic conductors, and a contribution due to the movement of positively charged *holes*. We then discuss *doped* semiconductors, introducing *N-type* and *P-type* semiconductors.

Our real interest will be in the description of two widely used devices constructed from N- and P-type semiconductors, namely the diode and the transistor.

In this chapter we restrict our discussion to the basic operation of the diode and show its use in some simple circuits. In particular, we will describe the use of the diode to rectify a sinusoidal e.m.f. Transistor operation will be discussed in Chapter 15.

## 14.1 Intrinsic semiconductors

Semiconductors are materials whose electrical resistivities lie between those of metallic conductors and good insulators. The resistivities of metallic conductors are of the order $10^{-8}$ $\Omega$m, whereas semiconductors have resistivities of order $10^{-2}$ to $10^2 \Omega$m. These values compare with the resistivities of good insulators, which are of the order $10^8$ $\Omega$m.

The most common material used in semiconductor devices is silicon (Si). We note that the atomic number of silicon is 14, so that the nucleus contains 14 protons. Since the nucleus consists of 14 protons, it follows that the silicon atom must have 14 electrons (recall that in their normal state atoms are electrically neutral). As noted in Chapter 2, the electrons orbit the nucleus in certain shells. Of great interest to us will be the role of the four valence electrons in the outermost shell when we discuss the crystalline structure of silicon.

Figure 14.1 shows the Bohr model of silicon. The figure shows the atomic core surrounded by the four valence electrons. The atomic core consists of the nucleus

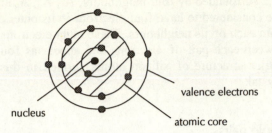

*Figure 14.1* **The silicon atom. The electrons in the outer shell are called the valence electrons. The atomic core consists of the nucleus and the electrons in the innermost shells**

(of charge $+14e$) surrounded by ten orbiting electrons (with total charge $-10e$), and has an overall charge of $+4e$, where $e$ is the magnitude of the charge on an electron. Recall our discussion in Section 2.3.

## The crystalline structure of semiconductors

To proceed we have to discuss how groups of silicon atoms join together to form a solid. This grouping of the atoms to form a solid is referred to as the crystalline structure. Figure 14.2 shows a two-dimensional representation of the structure of a silicon crystal.

Each of the valence electrons of an atom acts as if it pairs up, i.e. forms a bond, with a valence electron of an adjacent atom. For example, the central atom marked

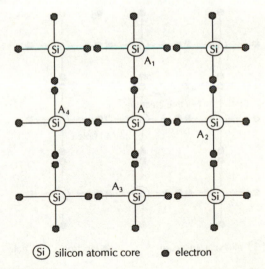

Si silicon atomic core    ● electron

*Figure 14.2* **The silicon lattice**

'A' in Figure is 14.2 surrounded by four neighbours, $A_1$, $A_2$, $A_3$ and $A_4$, and each of these atoms may be considered to have four electrons in its outer orbital and to share four more, one from each of its neighbours. These eight electrons are said to form a covalent bond between each pair of electrons and appear as four pairs around the atomic core. A lattice structure of silicon atoms formed in this way is called an *intrinsic* silicon crystal.

### Electron–hole pairs

At very low temperatures all the electrons are attached to their respective atomic cores and are not free to take part in the conduction of an electric current. Therefore the semiconductor material is an insulator. At room temperature, however, some of the electrons are able to acquire sufficient energy from the vibration of the lattice that they are able to break the covalent bond and move freely throughout the lattice. Such electrons are referred to as free, and if electric forces are applied they contribute to the electric current.

However, there is another interesting process which we must consider. Whenever an electron leaves its position in the lattice a vacancy exists which is known as a hole. Thus, every time an electron is freed, a hole is simultaneously created. This process is known as

### (thermal) electron–hole pair generation

See Figure 14.3. Note that for every free electron in the silicon lattice there is also a hole, i.e. free electrons and holes are generated in pairs.

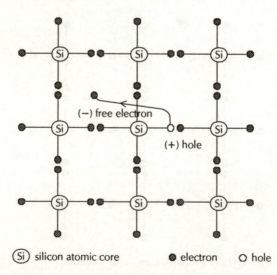

(Si) silicon atomic core   ● electron   ○ hole

*Figure 14.3* **Thermal electron–hole pair generation**

The negative charge associated with an electron is a familiar concept. By analogy, we associate a positive charge with the hole. This concept arises since, initially, the atomic core A must have a charge of $+4e$ units balanced by its four electrons with their total charge of $-4e$. After one of the electrons is freed the total charge of the atomic core and remaining valence electrons is

$$+4e - 3e = +1e$$

It is conventional to associate this charge with the hole and to use the concept that the hole has a positive charge $+e$ in the same manner that the electron has a negative charge $-e$.

## Charge flow in an intrinsic semiconductor

Let us now discuss the possibility that these holes are able to contribute to the flow of an electric current. We have argued that there are free electrons found in the silicon lattice because of thermal vibrations. It is quite possible for an electron from an atom, nearby where the hole has been formed, to escape from its position in the lattice and occupy the hole position. Consider Figure 14.4, which shows an electron at $A_2$ moving to the hole at $A_1$. This movement has two effects:

- The charge on the electron cancels the charge associated with the hole at $A_1$. The atom at this position originally had charge $+1e$ is now left in a neutral state.
- The atom at $A_2$ was originally uncharged. Now that the electron has left the atom it has a charge of $+e$.

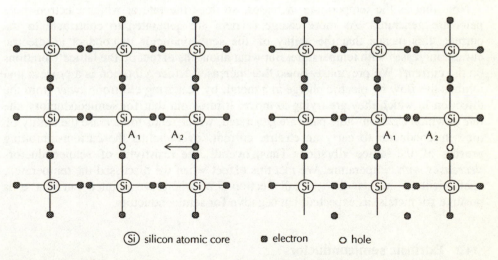

 silicon atomic core   ● electron   ○ hole

*Figure 14.4*   **Hole movement in a semiconductor: The electron moves from $A_2$ to $A_1$. The hole moves in the opposite direction**

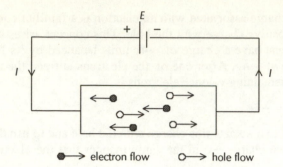

*Figure 14.5* **Charge movement in an intrinsic conductor**

It is very important to think of this in terms of the hole moving from atom $A_1$ to the atom $A_2$. This suggests that the hole is free to move through the lattice in the same manner that the free electron is able to move through the lattice.

It follows that if we apply electric forces to the silicon crystal the movement of the holes will contribute to the electric current; see Figure 14.5. At all non-zero temperatures charge will exist in the intrinsic crystal if an e.m.f. is applied across the semiconductor, as shown in the figure. The current involves the movement of negatively charged free electrons and positively charged holes. The electrons, having negative charge, will be attracted to the positive terminal, while the holes, having positive charge, will be attracted to the negative terminal. Recall that the current direction is defined as the direction of movement of positive charge and it follows that both the hole and the electron movement contribute to the current from the positive terminal of the source to the negative terminal.

Note that as the temperature increases, so does the rate at which electron–hole pairs are generated. As more charge carriers are generated to contribute to the current, this means that the ability of the semiconductor to conduct an electric current increases with temperature. But what about the effect of the lattice vibrations on the current? We previously noted that increased lattice vibration is a process that inhibits the flow of electric charge in a metal, by scattering electrons away from the direction in which they are trying to move. It turns out that for semiconductors, the rate at which electron–hole pairs are generated, which tends to increase the ability of the semiconductor to carry an electric current, overwhelms the current-inhibiting process of the lattice vibrations Thus, overall, the resistivity of semiconductors *decreases* with temperature. We met this effect when we discussed the temperature coefficient of resistance $\alpha$; recall Section 5.2, where we pointed out that $\alpha$ is positive for metals, as expected, but negative for semiconductors.

## 14.2 Extrinsic semiconductors

Although the ability of a semiconductor to conduct an electric current increases with temperature, much larger increases are effected by introducing to the pure material selected impurities, often much less than one impurity atom per million intrinsic atoms. This process of adding selected impurities to alter the ability of the semiconductor to conduct an electric current is referred to as

*doping*

and the semiconductor is said to be

*extrinsic* or *doped*

We discuss the two types of doping to produce N-type and P-type semiconductors. We will find that in *N-type* material the *majority* charge carriers are *electrons*, while in *P-type* material the majority carriers are *holes*.

## N-type semiconductors

Certain elements, such as phosphorus (P) or arsenic (As) have five electrons in the valence shell, compared with the four associated with the silicon atom. When these elements are introduced as impurities into the silicon crystal, four of the five electrons form the covalent bonds discussed earlier and fit into the lattice. The fifth electron remains only very loosely attached to the parent impurity atom and only a very small amount of energy is needed to release it. At room temperature almost all of these loosely held electrons belonging to the impurity atoms are free to move in the lattice.

An impurity atom which contributes free electrons to the lattice is called a *donor* impurity. The semiconductor now has an excess of electrons or negative charges and is known as an N-type semiconductor. Figure 14.6 shows the fifth electron initially

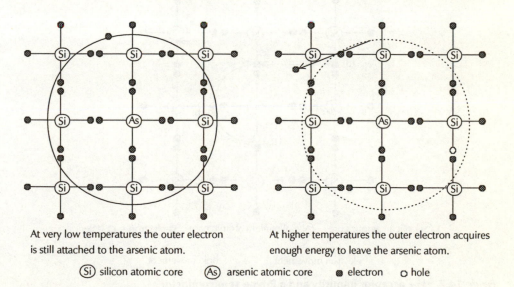

At very low temperatures the outer electron is still attached to the arsenic atom.

At higher temperatures the outer electron acquires enough energy to leave the arsenic atom.

(Si) silicon atomic core    (As) arsenic atomic core    ● electron    ○ hole

*Figure 14.6*  **The donor impurity and an N-type semiconductor**

loosely attached to an arsenic atomic core. At room temperature it is able to acquire energy from the vibrating lattice to leave the atomic core, and move from A to B.

It should be noted in this instance that the existence of the free electron does not produce a hole. It does, however, leave an *immobile* arsenic impurity core in the lattice with a net positive charge. Being very massive, the arsenic core does not respond readily to electric forces and cannot, therefore, contribute to an electric current.

Also note that there will be holes present due to thermal generation, as discussed previously. Usually the doping is such that the number of holes is very much smaller than the number of free electrons. For N-type material, electrons are the majority carriers, whereas holes are minority carriers.

## P-type semiconductors

Impurities such as boron (B) or aluminium (Al) can also be added to the pure semiconductor. These atoms have only three electrons in the valence shell in place of the four of the silicon atom. Thus there is a vacancy for an electron associated with each impurity atom. The vacancy can be filled by an electron from a nearby atom creating a net positive charge, i.e. a hole, there.

Since this type of impurity accepts electrons it is known as an *acceptor* impurity and the semiconductor is known as a P-type semiconductor. The impurity atom itself acquires a net negative charge from the electron, but since the ion is immobile at the lattice site, cannot contribute to the conduction of an electric current. Thus, the creation of the hole does not create a corresponding free electron.

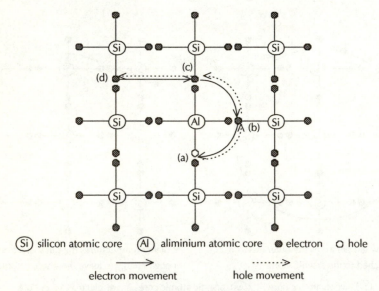

*Figure 14.7*  **The acceptor impurity and a P-type semiconductor**

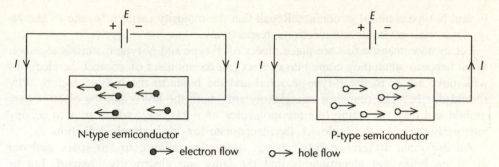

*Figure 14.8* **Charge movement in N- and P-type semiconductors**

Consider Figure 14.7. The hole shown at (a) is free to move throughout the lattice in the manner indicated – the electron from site (b) moves into position (a), leaving a hole at (b). The electron from position (c) then moves into (b). Alternatively, we can argue that the hole at (b) moves to (c). The electron from (d) moves into position (c), which is the same as the hole from (c) moving to (d). Thus, the hole seems to move through the lattice.

Again, note that there will be free electrons present due to thermal generation, as discussed previously. Usually the doping is such that the number of free electrons is very much smaller than the number of holes. For P-type material, holes are the majority carriers, whereas electrons are minority carriers.

Figure 14.8 summarizes this discussion. Strictly, Figure 14.8 shows only the contribution to the current from the *majority* charge carriers. As noted above, in N-type material there are present a small amount of holes which would contribute to the current, while in P-type material there are present a small amount of electrons which would contribute to the current there.

## 14.3 The diode

In this section we discuss the junction that is formed when pieces of N-type and P-type materials are joined together. This forms the basis of the diode. We then proceed to describe what happens when a battery is placed across such an arrangement. We will find that the size of current depends upon how the battery is placed across the diode. This will lead us to introduce the terms

*forward* and *reverse biasing*

After this we will describe some simple uses of the diode.

### The PN junction

The simplest semiconductor device to consider is the PN junction, which consists of

P- and N-type material in contact. Recall that the majority carriers for the P- and N-type materials are holes and electrons, respectively.

Let us now imagine that we place pieces of P-type and N-type materials together. What happens when they come into contact? At the moment of contact the electrons will move across to the P-type material and the holes to the N-type region. Why should this be? The reason is straightforward – both the holes and the electrons are mobile and they are acting like the molecules of a gas, say, which tries to occupy uniformly the space available, i.e. the mechanism for this is simple diffusion.

An important difference between gas molecules moving to fill space and our diffusing holes and electrons, is that the latter are electrically charged. Let us consider the effect of the electrons diffusing across to the P-type side. These electrons *recombine* with the holes there and neutralize them, leaving an excess of (immobile) negatively charged acceptor atoms. Similarly, as the holes move from the P-type to the N-type side, they recombine with the electrons, leaving (immobile) positively charged donor atoms behind. Therefore it becomes increasingly difficult for holes and electrons to cross over, because this charge imbalance inhibits further hole and electron movement across the junction. This charge imbalance increases until an equilibrium is reached.

Once equilibrium has been established a thin layer either side of the junction is formed, typically of the order of a few $\mu$m, which consists of immobile acceptor atomic cores on the P-type side and immobile donor atomic cores on the N-type side; see Figure 14.9. This region has very few mobile charge carriers present and is referred to as

the *depletion region* or the *depletion layer*

This terminology accurately describes the region since it has been depleted of the mobile charge carriers originally present. Since the depletion layer has very few mobile charge carriers it represents a region of the diode of high resistance.

As suggested, an excess negative charge is developed on the P-type side while an excess positive charge is developed on the N-type side. Because of this there is a p.d.

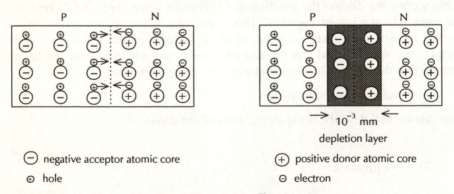

$\ominus$ negative acceptor atomic core      $\oplus$ positive donor atomic core

$\oplus$ hole      $\ominus$ electron

*Figure 14.9* **The PN junction and the diffusion of holes and electrons**

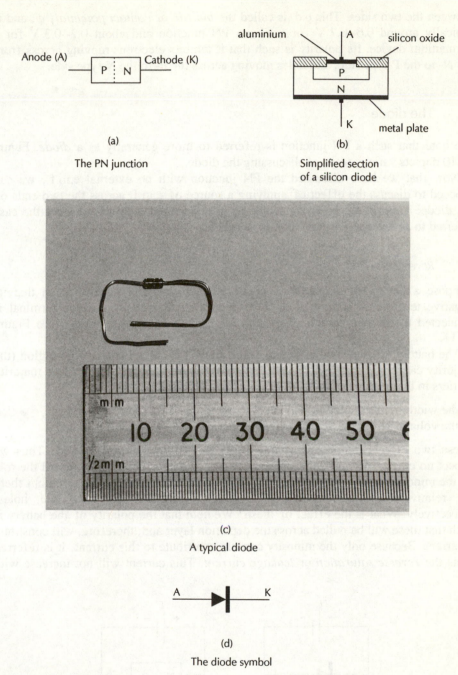

(a)

The PN junction

(b)

Simplified section
of a silicon diode

(c)

A typical diode

(d)

The diode symbol

*Figure 14.10*    **The diode**

between the two sides. This p.d. is called the *built-in* or *contact potential*, $\phi$, and is typically around 0.6–0.7 V for a silicon PN junction and about 0.2–0.3 V for a germanium device. Its polarity is such that it inhibits electrons moving across from the N- to the P-type side, and holes moving across the P- to the N-type side.

## The diode

We note that such a PN junction is referred to more generally as a *diode*. Figure 14.10 depicts various ways of discussing the diode.

Now that we have discussed the PN junction with no external e.m.f., we can proceed to discuss the effect of applying a source of e.m.f. across the two ends of the diode. This should be straightforward to understand. We first discuss the case referred to as reverse bias, and then forward bias.

### Reverse bias

Suppose a battery of e.m.f. $E$ is connected across a PN junction such that its negative terminal is connected to the P-type material and its positive terminal is connected to the N-type side. The diode is said to be reverse biased; see Figure 14.11.

The battery as connected in Figure 14.11 draws the holes in the P-type region (the majority carriers in that region) and the electrons in the N-type region (the majority carriers in that region) away from the junction. This has two effects:

- the width of the depletion region is increased (see Figure 14.12), and
- the voltage across the depletion region increases.

These two effects inhibit any charge flow due to the majority carriers. Thus we expect no current through the circuit. However, we have not yet mentioned the role of the minority carriers in each region. Recall that in the P- and N-type regions there are relatively small numbers of thermally generated electrons and holes, respectively. What is the effect of these? We note that the polarity of the battery is such that these will be pulled across the depletion layer and, therefore, will constitute a current. Because only the minority carriers contribute to this current, it is referred to as the *reverse saturation* or *leakage current*. This current will not increase with

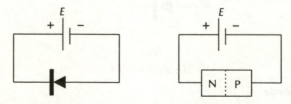

*Figure 14.11*  **Reverse biasing the diode. The current flow is zero**

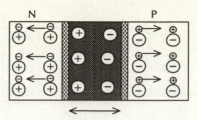

depletion layer with no biasing

*Figure 14.12* **No contribution to the current from the majority carriers. Note the effect of the reverse bias on the width of the depletion layer (shown by the cross hatching)**

the supply voltage once the majority carrier current has been eliminated. Recall that the minority carriers are thermally generated so that the reverse saturation current is temperature dependent. For a low-power silicon diode at room temperature this current has a magnitude of about 20 nA, while for germanium, another semiconductor material, the leakage current is of the order of 1–2 μA.

### Forward bias

Suppose now the battery of e.m.f. $E$ is connected across a PN junction such that its positive terminal is connected to the P-type material and its negative terminal is connected to the N-type side. This diode is said to be forward biased and is shown in Figure 14.13. The effect of the forward bias is to push majority carriers towards the junction, which has two major consequences:

- the width of the depletion region is reduced (see Figure 14.14), and
- the voltage across the depletion region is reduced.

These two effects encourage a charge flow due to the majority carriers. Thus, we expect a current through the circuit. But we must take into account the contact potential, $\phi$, present between the junction. Recall that the contact potential is due to excess negative charge on the P-type side of the junction and excess positive charge on the N-type side. The polarity of the contact potential is such that electrons moving from the N-type side to the P-type side, and holes moving from the P-type side to the N-type side, require energy to do so. This energy is supplied by the

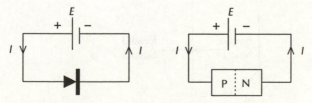

*Figure 14.13* **Forward biasing the diode. The current flow is large**

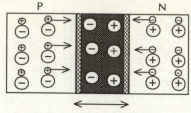

depletion layer with no biasing

*Figure 14.14*  **The contribution to the current from the majority carriers. Note the effect of forward biasing on the width of the depletion layer (shown by the cross hatching)**

battery. It follows that some of the energy supplied by the battery must be used in moving electrons and holes across the junction. In some sense we can regard this energy as 'lost'. Thus, when the diode is forward biased, we only expect a current if the e.m.f. of the battery is greater then the contact potential, $\phi$. If the e.m.f. is smaller than the contact potential, $\phi$, then we only expect a negligible current.

■ **COMMENT**

■ The symbol of the diode indicates the direction in which the conventional current flows through the device; see Figure 14.10(d).

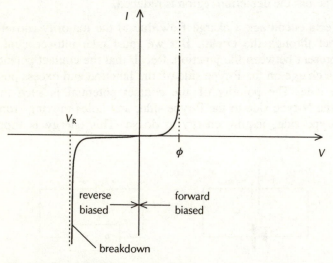

*Figure 14.15*  **The I–V characteristics of the diode**

### The I–V characteristics for the diode

Based upon our previous discussions, we now present the *I–V* characteristics for the diode; see Figure 14.15. Recall from Section 5.1 that the *I–V* characteristics just tell us how the current through the component varies with the potential difference across it.

From Figure 14.15 we note that if the diode is reverse biased, the current through the circuit is roughly constant and small, being due to minority charge carriers. As the e.m.f. increases from zero the current starts to increase. For e.m.f.s smaller than the contact potential the current is still small. As the e.m.f. increases above the contact potential the current increases more rapidly. We will discuss the breakdown phenomenon later.

■■■ **WORKED EXAMPLE 14.1**

A particular diode has the *I–V* characteristic curve shown in Figure 14.16. Determine the resistances at points A, B, C and D.

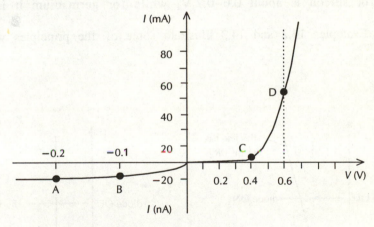

*Figure 14.16* **The *I–V* characteristics of the diode**

– The resistances at points A–D are given in Table 14.1. Note the change of scales on the negative axes of the graph.

*Table 14.1*

| Point | p.d. (V) | I | R |
|-------|----------|---------|---------|
| A | −0.2 | −20 nA | 10 MΩ |
| B | −0.1 | −18 nA | 5.6 MΩ |
| C | 0.4 | 5 mA | 80 Ω |
| D | 0.6 | 55 mA | 10.9 Ω |

Note that from Section 5.1 the diode is an example of a non-Ohmic conductor, since its *I–V* characteristics are not linear.

Worked Example 14.1 is a re-working of Worked Example 5.2. The only difference is that the current ranges have been suitably modified.

This example suggests two convenient approximations for this curve, which are presented in Figure 14.17. In Figure 14.17(a) the diode is approximated as a device with two states, 'OFF' and 'ON'. The OFF state has no current when the p.d. across the diode is less than the contact potential, $\phi$, and an ON state when the p.d. across the diode is greater than the contact potential, where the current rises linearly as the e.m.f. rises above $\phi$. Therefore the diode can be represented as an open circuit (or very high resistance) if $V < \phi$, and as a resistance $R_d$ if $V > \phi$. Typically the forward resistance $R_d$ has a value of about 10 $\Omega$.

Figure 14.17(b) shows a second approximation where the forward resistance $R_d$ can be ignored. In this approximation when the diode is ON, i.e. when it conducts, the p.d. across the diode is at a constant value given by the value of the contact potential. When the diode is ON it can be represented by an equivalent circuit, as depicted in Figure 14.18. Typically we find the contact potential for silicon is about 0.6–0.7 V, while for germanium it is about 0.2–0.3 V.

Worked Examples 14.2 and 14.3 illustrate some of the principles we have discussed.

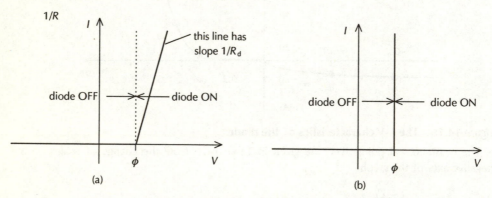

Figure 14.17   **Two convenient approximations for the *I–V* characteristics of the diode**

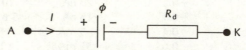

Figure 14.18   **Equivalent circuit of PN diode when ON**

━━━ **WORKED EXAMPLE 14.2**

Determine the current through the circuit in Figure 14.19 if the diode has a contact potential $\phi = 0.6$ V and a forward resistance of 5 Ω.

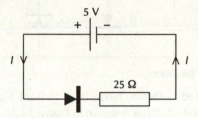

*Figure 14.19* **Determine the current** *I*

– Since the diode in Figure 14.19 is forward biased, we can replace it with the equivalent circuit depicted in Figure 14.18. Therefore, we are led to consider the circuit shown in Figure 14.20.

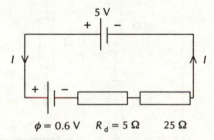

*Figure 14.20* **The equivalent circuit**

This circuit is straightforward to solve. We find that the current through the circuit, *I*, is

$$I = \frac{(5 - 0.6)\text{V}}{(25 + 5)\Omega} = 0.147 \text{ A}$$

Worked Example 14.3 involves two diodes:

━━━ **WORKED EXAMPLE 14.3**

The two diodes in Figure 14.21 can be assumed to have a forward resistance of 5 Ω and an infinite reverse resistance. Assume that the contact potential is negligible compared to any p.d. across the components and thus can be ignored. Determine the current through each branch of the circuit if the battery is connected, as shown in the figure, and when the battery is reversed.

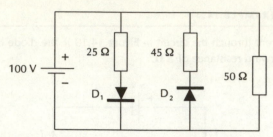

*Figure 14.21* **A two-diode problem**

– With the battery connected as shown, diode $D_1$ is forward biased and diode $D_2$ is reverse biased. Therefore, the equivalent circuit is shown in Figure 14.22. This follows since the diode $D_1$ is ON and has an effective resistance of 5 $\Omega$, while the diode $D_2$ is OFF and is effectively an open circuit. This problem has now been reduced to a simple resistor network problem. It is straightforward to determine the current through each branch of the circuit. We find

$$I_{(25\,\Omega)} = 3.333 \text{ A}, \qquad I_{(45\,\Omega)} = 0, \qquad I_{(50\,\Omega)} = 2 \text{ A}$$

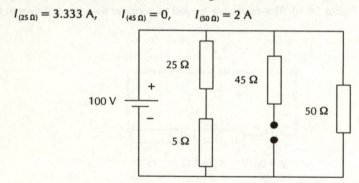

*Figure 14.22* **The diode $D_1$ is forward biased**

When the battery is reversed the diode $D_1$ is now reverse biased and hence does not allow a current to pass, while the diode $D_2$ is now forward biased and, in our approximation, acts like a resistor of 5 $\Omega$. The equivalent circuit is shown in Figure 14.23. It is straightforward to determine the current through each resistor. We find

$$I_{(25\,\Omega)} = 0, \qquad I_{(45\,\Omega)} = 2 \text{ A}, \qquad I_{(50\,\Omega)} = 2 \text{ A}$$

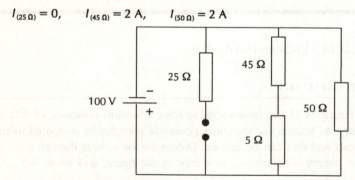

*Figure 14.23* **The diode $D_2$ is forward biased**

### Breakdown

In Figure 14.15 we notice that there is a reverse bias voltage, $V_R$, such that a slight increase beyond this value produces a very large increase in the current through it. The diode is said to have broken down. In practical circuits, this breakdown voltage is an important parameter of the diode and will be quoted by the manufacturer. There are two mechanisms which cause breakdown:

- *Avalanche breakdown:* Under reverse bias conditions the current is due to the flow of minority carriers through the depletion layer. The polarity of the reverse bias voltage is such that the minority carriers gain energy as they move across the depletion layer. If the reverse bias voltage is large enough these minority carriers gain enough energy to create further electron–hole pairs by colliding with atoms in the depletion layer. These will also be accelerated by the reverse bias voltage and in turn may generate more electron–hole pairs. Since the number of charge carriers is increasing, the current increases.
- *Zener breakdown:* Zener breakdown depends on the fact that if the doping concentration is high the depletion layer will be narrow. The mechanism of Zener breakdown is complicated and will not be discussed here.

### 14.4 The diode as a rectifier

From our discussion we note that the diode presents a small resistance to the current for one polarity of the e.m.f. and a high resistance for the opposite polarity. This is used to great effect in the

*rectifier circuit*

Consider Figure 14.24(a), which shows a simple rectifier circuit consisting of a diode connected in series with a sinusoidal source of e.m.f., $\mathscr{E}$. For one half of the cycle the point P will be positive with respect to the point Q. In this case the diode is forward biased so that it will conduct. In this case the current will flow through the

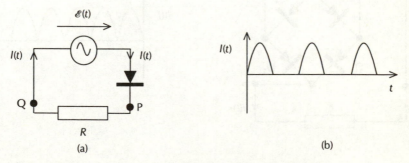

*Figure 14.24*   **A simple rectifier circuit. (a) The circuit; (b) the output**

load resistor $R$ from the point P to the point Q. The current will be in phase with the source of the e.m.f. since there are no inductors or capacitors in the circuit.

For the next half of the cycle the point P will be negative with respect to the point Q and, therefore, the diode will be reverse biased. The reverse biased diode acts like an open circuit so that no charge will flow through the resistor. It follows that the current through the resistor will only flow in one direction, as indicated in Figure 14.24(b). This current is referred to as unidirectional, since although the magnitude of the current depends upon time, the direction does not. The circuit in Figure 14.24 is referred as a

*half wave rectifier*

Figure 14.25(a) depicts another arrangement, the so-called

*bridge rectifier*

which is an example of a

*full wave rectifier*

The bridge rectifier consists of four diodes, $D_1$, $D_2$, $D_3$ and $D_4$, connected to the alternating supply in the manner indicated in the figure. For one half of the cycle the point P will be positive with respect to the point Q. In this case the diodes $D_1$ and $D_3$ are forward biased so that these diodes will conduct, while the diodes $D_2$ and $D_4$ are reverse biased and so act like open circuits. In this case the current will flow through the load resistor $R$ from the point A to the point B.

In the next half cycle the point P will be negative with respect to the point Q. It now follows that the diodes $D_2$ and $D_4$ are forward biased and, therefore, conduct.

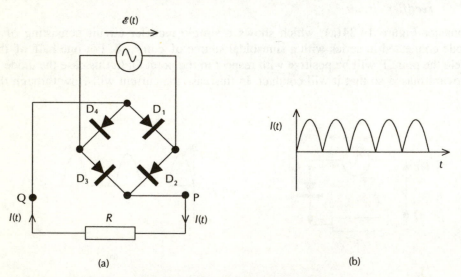

(a)             (b)

*Figure 14.25* **The bridge rectifier. (a) The circuit; (b) the output**

The diodes $D_1$ and $D_3$ are now reverse biased and act like open circuits. From the arrangement of the diodes it follows that the current through the load resistor is in the same direction as previously.

It follows that the current flows through the resistor in only one direction, irrespective of the relative polarities of points P and Q. The output is depicted in Figure 14.25(b).

## 14.5 Other uses of diodes

We now briefly a short discuss some other important uses of the diode.

### The Zener diode

In an ordinary diode if the reverse bias is increased until the depletion layer diode breaks down, the diode suffers permanent damage. A Zener diode is constructed to be used in the breakdown region so long as the current flowing through the device is limited. As the reverse voltage across the diode is increased the current remains small until the breakdown or Zener voltage is reached (see the $I–V$ characteristics for the diode; Figure 14.15). At this voltage the current through the diode suddenly increases. If the reverse bias falls below the Zener voltage, the current becomes negligible again. Thus, the voltage across the diode remains almost constant at the Zener voltage over a wide range of reverse currents. It is this property of the Zener diode that is utilized in voltage supply stabilizers.

### The light-emitting diode (LED)

When a PN junction is forward biased electrons move across from the N- to the P-side, where they recombine with holes near the junction. Similarly, holes move across the junction from the P- to the N-side, where they recombine with electrons. As recombination takes place, energy is released. Usually this energy release goes into heating the diode. In gallium arsenide phosphide the energy is emitted as light which is able to leave the diode if the junction is formed very close to the surface of the material. Typical uses include calculator displays.

### The photodiode

The photodiode consists of a PN junction in a case with a transparent 'window' through which light can enter. A photodiode is operated in reverse bias and the leakage (minority carrier) current increases as the amount of light falling on the junction increases. This is due to the incoming light forming extra electron–hole pairs (i.e. the reverse effect of electron–hole recombination). Typical uses of the photodiode are as fast counters which generate a current pulse every time a beam of light is interrupted.

## The varicap (varactor) diode

We have noted that when the diode is reverse biased the depletion layer acts like a very large resistance sandwiched between two low-resistance P- and N-type materials (see Figure 14.12). Let us rephrase this statement: the diode acts as an insulator sandwiched between two conductors. But this is just a description of the capacitor. Thus, we see that a reverse biased diode can behave as a capacitor. The effective capacitance depends upon the area of the PN junction and the thickness of the depletion layer. Since the width of the depletion layer depends on the reverse voltage we have a useful mechanism of changing the capacitance. Typically, a varicap diode has a range of capacitance of about 2–10 pF.

## 14.6 Summary

We presented a brief discussion of charge movement in a class of materials referred to as semiconductors, first discussing an intrinsic semiconductor and then the doped semiconductor. We found that there are two types of doped material, namely N- and P-type material, in which the majority carriers in each material are electrons and holes, respectively.

   Our real interest lay in the description of a widely used device constructed from N- and P-type semiconductors, namely the diode. After a discussion of the basic operation of the PN junction, i.e. the diode, we showed how the diode acts as a rectifier, offering a low resistance when a battery is connected across it in one direction, and offering a high resistance when the polarity is reversed.

   When forward biased an appreciable current can flow if the p.d. across the diode is larger than the contact potential. In such a situation the resistance of the diode is small. When reverse biased the depletion layer widens and the resistance of the device is very large and only a small current flows.

## ▬▬ Problems

**14.1**   Figure P14.1 shows a diode in series with a 15 Ω resistor and a battery of e.m.f. 10 V.
   (a)   Determine the current through the following circuit if the diode has a contact potential $\phi = 0.6$ V and a forward resistance of 5 Ω.
   (b)   Repeat the question if the contact $\phi = 0$ V.
   (c)   What is the current through the circuit if the battery is reversed?
   (d)   Repeat part (a) if the battery is replaced by the alternating supply

   $$\mathscr{E} = 5 \sin(200\pi t) \text{ V}$$

   i.e. determine the current through the circuit over one complete cycle.

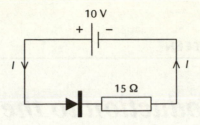

*Figure P14.1*

> *Hint:* First determine the current taking the contact potential $\phi = 0$ V, then consider the case when $\phi = 0.6$ V. When the p.d. across the diode is less than the contact potential the current through the diode can be taken as zero.

**14.2** The two diodes in Figure P14.2 can be assumed to have a forward resistance of 5 Ω and an infinite reverse resistance.

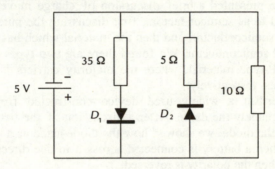

*Figure P14.2*

> (a) Determine the current through each branch of the circuit if the battery is connected as shown in the figure if the contact potential of each diode can be ignored.
> (b) Repeat the question if the battery is reversed.
> (c) Repeat part (a) if each diode has a contact potential of 0.6 V.

○ **14.3** Determine the frequency and average voltage for the output of a half and full wave rectifier if the input is a sinusoidal wave of peak amplitude 10 V and frequency 500 Hz.
> *Hint:* Make use of the results of Chapter 12.

# CHAPTER FIFTEEN

# An introduction to the transistor

In Chapter 14 we presented a brief discussion of charge movement in a class of materials referred to as semiconductors, first discussing the pure material, referred to as an intrinsic semiconductor, and then the material which has selected impurities added, the doped semiconductor. We found there are two types of doped material, namely N- and P-type material, where the majority carriers in each material are electrons and holes, respectively.

We then described a widely used device constructed from N- and P-type semiconductors, namely the diode. After a discussion of the basic operation of the PN junction, i.e. the diode, we showed how the diode acted as a rectifier, offering a low resistance when a battery is connected across it in one direction and offering a high resistance when the polarity is reversed.

In this chapter we discuss another semiconductor device, the transistor. Many types of transistor are available, but they can be effectively grouped into two types:

■ The bipolar junction transistor (BJT). The term *bipolar* refers to the fact that its operation depends on the flow of both electrons and holes. Recall that the diode is made of a 'sandwich' of two doped semiconductor materials, namely N- and P-type. The next obvious arrangement of N- and P-type semiconductor materials is an NPN or PNP 'sandwich'. This is the arrangement of the BJT.

■ The unipolar or field effect transistor (FET). The term *unipolar* refers to the fact that transistor action depends on the flow of only one type of charge carrier for its operation (i.e. either electrons or holes). This type of transistor is just an arrangement of P-type and N-type material, but which is different from the arrangement in the diode.

In practice, the transistor is a three-terminal device in which the current flowing through two of the terminals is controlled by a signal on the third terminal. It is this property which enables the transistor to be used as a switch and an amplifier.

Since transistors are capable of amplifying signals they are termed *active* components. This is to distinguish them from components such as resistors, inductors, capacitors and diodes, which are not able to amplify signals, and which are referred to as *passive* components. We refer to the study and application of transistor circuits as *electronics*. The distinction or difference between *electricity* and *electronics* is sometimes not clear or apparent in the literature – both subjects are essentially discussions of the movement of electric charge. We are essentially arguing that electronics is the study of active electrical components. This is not a universal definition; some authors, for example, introduce the term electronics when discussing semiconductors or the diode.

In this chapter we will first discuss the BJT, describing the basic operation of such a device and then two important uses of the transistor, namely as a switch and as an amplifier. We will find that the BJT is a current controlled device. After this, we then go on to describe the FET. We will note that the FET is a voltage controlled device.

## 15.1  The bipolar junction transistor (BJT)

We now describe the operation of the bipolar junction transistor (BJT). The BJT is a three-layer semiconductor device that has three terminals, the *base* (B), the *collector* (C) and the *emitter* (E). There are two types of transistor: the NPN and the PNP transistor. The NPN type is made of a narrow layer of P-type material sandwiched between two N-type layers, while the PNP type is made of a narrow layer of N-type material sandwiched between two P-type layers. In the following we discuss only the NPN arrangement.

As just mentioned, the NPN transistor consists of a narrow layer of P-type material sandwiched between two N-type layers as shown in Figure 15.1(a), while Figure 15.1(b) shows a simplified section of a transistor. Many aspects of transistor action can be understood if we think of the transistor as a pair of diodes placed back-to-back, as depicted in Figure 15.1(c). A typical transistor is shown in Figure 15.1(d), and the accepted symbol is shown in Figure 15.1(e).

As discussed, the three layers are called the emitter (E), base (B) and collector (C). The emitter is differentiated from the collector in Figure 15.1(e) by the use of an arrow. This terminology and the significance of the direction of this arrow will be explained shortly.

### Principle of operation

Two important features of the BJT are that the emitter is much more heavily doped than the base, and the base region is very thin.

Let us consider the BJT in its normal operating condition; see Figure 15.2. Note that the collector is at a higher potential than the emitter, i.e. $V_{CE} > 0$. Note that in Figure 15.2 the base–emitter junction is forward biased by the battery $V_{BE}$ so that this PN junction will conduct as an ordinary diode. Most of the current will consist of electrons crossing from the emitter into the base since the emitter region is more heavily doped than the base region.

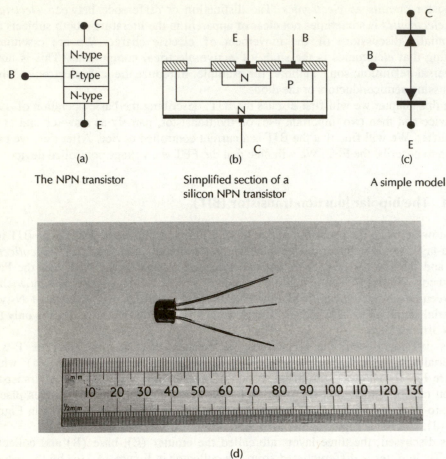

(a)

The NPN transistor

(b)

Simplified section of a
silicon NPN transistor

(c)

A simple model

(d)

A typical transistor

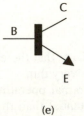

(e)

Symbol for NPN transistor

*Figure 15.1* **The NPN transistor**

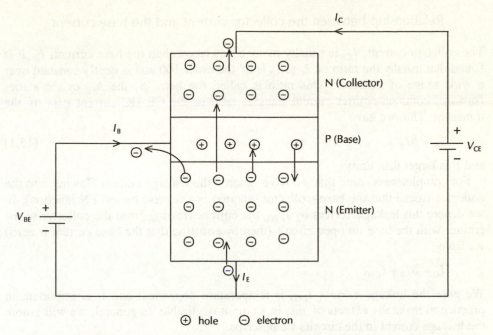

*Figure 15.2* **Charge movement and current flow in the NPN bipolar junction transistor**

Since the base is made to be very thin, typically of the order of a few $\mu$m, once these electrons have been injected into the base, where they are minority carriers (recall the discussion in Section 14.2), they are in the vicinity of the base–collector junction and are swept across into the collector. This follows since the base–collector junction is reverse biased. These electrons form the collector current, $I_C$.

There are several factors which cause the number of charge carriers crossing from the base to the collector to be less than the number crossing the base–emitter junction:

- The base is a P-type region, so holes pass from the base to the emitter in the usual manner for a forward biased PN junction. However, because the base region is only lightly doped compared with the emitter this will only be a small effect.
- Some of the electrons entering the base will recombine with holes before reaching the base–collector junction. Again, because the base region is lightly doped and is thin this will be a small effect.
- Some of the electrons pass from the emitter to the base. If the base region is shaped this effect will be small.

Therefore we expect the collector current, $I_C$, to be very nearly equal to the emitter current, $I_E$.

## Relationship between the collector current and the base current

The collector current, $I_C$, is usually many times larger than the base current, $I_B$. It is found that usually the ratio of $I_C$ to $I_B$ is of the order 100 and is nearly constant over a wide range of currents. This ratio is called the beta, $\beta$, the $h_{FE}$ or the static, forward, common-emitter current transfer ratio or the CE DC current gain of the transistor. Thus we have

$$I_C = \beta I_B \tag{15.1}$$

and $\beta$ is larger than unity.

For completeness, note that we have ignored the leakage current flowing into the collector (recall that the base–collector junction is a reverse biased PN junction). If we denote this leakage current by $I_{CEO}$, the current flowing from the collector to the emitter with the base on open circuit (thereby ensuring that the base current is zero) we have

$$I_C = \beta I_B + I_{CEO}$$

We note the leakage current $I_{CEO}$ is temperature dependent and it is important in practice to make the effects of this in a circuit negligible. In general, we will ignore the leakage current in the circuits we describe.

## Current conventions and the BJT symbol

Figure 15.3 shows the NPN BJT connected in the usual common-emitter configuration together with the usual current through the device. Note that according to Kirchhoff's current law we have

$$I_E = I_C + I_B \tag{15.2}$$

Substituting (15.1) into (15.2), we have that the emitter current $I_E$ is related to the base current (ignoring leakage) as

$$I_E = (1 + \beta)I_B \tag{15.3}$$

This gives us a clue as how the BJT acts as an amplifier. If the base current $I_B$

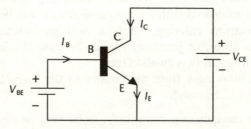

*Figure 15.3* **Current conventions in the NPN bipolar junction transistor**

changes by an amount $\Delta I_B$, then the emitter current $I_E$ changes by the amount $\Delta I_E$, given as

$$\Delta I_E = (1 + \beta) \Delta I_B \tag{15.4}$$

Since we have $\beta$ larger than unity, the change in the emitter current is larger than the change in the base current. Because the base current controls the emitter current we describe the BJT as a current-operated device.

The configuration presented above is termed the

*common emitter configuration* or *mode*

Clearly this is reasonable terminology since the emitter terminal is common to both input and output voltages.

### ■ COMMENTS

- For an NPN transistor the electrons flow from the emitter region into the base region and continue on to be collected in the collector region. Because conventional current flows in the opposite direction to electron flow, the arrow on the circuit symbol indicates the flow of conventional current through the device.
- The names of the terminals come from the parts they play in the operation of the device:
  - the emitter *emits* electrons;
  - the collector *collects* electrons;
  - the base *controls* the flow of the electrons from the emitter to the collector.
- The fact that a transistor is made up of a semiconductor sandwich of N- and P-type materials means that there are two types of charge carriers to be considered in the charge movement within the device, hence the term *bipolar* in the expression BJT. However, to understand the simple operation of the device it is only necessary to consider one of the charge carriers – electrons for the NPN device and holes for the PNP device.
- For completeness, we note that arguments presented above can be repeated for the PNP transistor if we replace 'hole' by 'electron' and reverse all the batteries in Figure 15.2. We also point out that the symbol for the PNP transistor is similar to that for the NPN transistor given in Figure 15.1(d) except that the arrow on the emitter is in the opposite direction. From our discussion of the significance of the direction of the arrow on the emitter in the NPN case, this is reasonable.

### The common-emitter (CE) characteristics

We now discuss a particular arrangement of the BJT which is typically used in a great many circuits, the so-called common-emitter (CE) configuration.

We now describe in some detail the CE characteristics. We are interested in two types of data, namely, how the base current, $I_B$, depends upon the base–emitter p.d., $V_{BE}$, and how the collector current, $I_C$, depends upon the collector–emitter p.d., $V_{CE}$.

We will find that this characteristic depends upon the current through the base so that we will have an $I_C$ versus $V_{CE}$ plot for a given value of $I_B$.

## Input (base–emitter) characteristics $I_B$ versus $V_{BE}$

Consider the BJT connected as shown in Figure 15.4(a). This is very similar to Figure 15.3 except that the arrow through the battery $V_{BE}$ indicates a variable supply. By varying the p.d. across the base–emitter junction and measuring the current through the base for a fixed collector–emitter p.d. we obtain the CE input characteristics, as shown in Figure 15.4(b).

The input characteristic of the BJT relates the current flowing through the base to the p.d. between the base and the emitter. The CE input characteristic is closely related to that of a forward biased diode, although there are some subtle points we have to discuss:

- The characteristic is similar to that of a forward biased diode, appreciable current only flows through the base when the p.d. is about 0.6–0.7 V for a silicon device and 0.2–0.3 V for a germanium device.
- The characteristic is only weakly dependent on the supply $V_{CE}$. When the supply $V_{CE}$ is large the depletion layer at the base–collector junction is wider than when $V_{CE}$ is small. As the depletion layer at the base–collector region gets wider, this makes the effective base region smaller. If the base region is smaller there will be less recombination of holes and electrons, and therefore we expect more electrons from the emitter region to pass through to the collector region and less current through the base. However, in practice, the dependence of $I_B$ on $V_{CE}$ is quite weak and is often ignored.
- Note that in the region marked A the variation of the current with $V_{CE}$ is nearly linear.

The BJT is normally operated with the base–emitter forward biased. We note that sometimes it is operated as a reverse biased junction. If this is so we must be careful that in this case the voltage does not exceed a value such that breakdown occurs.

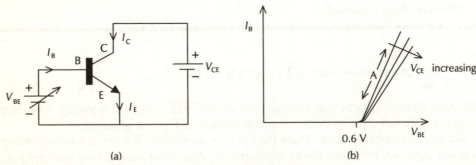

(a)                                    (b)

*Figure 15.4*  **The CE configuration and input characteristics**

Recall, for a reverse biased PN junction breakdown is the rapid increase of current as the voltage is increased beyond a certain value. Manufacturers usually quote the maximum allowed reverse bias voltage for the device before this breakdown occurs.

### Output (collector) characteristics $I_C$ versus $V_{CE}$

We now discuss how the collector current, $I_C$, depends upon the p.d. between the collector–emitter, $V_{CE}$. For this we consider the BJT connected as shown in Figure 15.5(a). Again, this is very similar to Figure 15.3 except now we have connected a variable resistor $R_B$ between the base–emitter supply and the base. Since we wish to determine how the collector current varies with the supply $V_{CE}$ we make use of a variable supply for $V_{CE}$. By varying the resistance of $R_B$ we can ensure a constant base current. By varying the p.d. across the collector–emitter and measuring the current through the collector, we obtain the CE output characteristics, as shown in Figure 15.5(b).

We first note that with no current through the base, i.e. $I_B = 0$, a small current flows through the collector. This can be identified as the leakage current, $I_{CEO}$ (usually negligible) and is shown in Figure 15.5(b) as the horizontal line almost coincident with the $V_{CE}$ axis.

Now let us increase the base current to some fixed small value, for example about 10 µA. We find for small values of $V_{CE}$, typically $V_{CE}$ less than about 0.2 V, the collector current rises sharply with $V_{CE}$. In this region the BJT is said to be operating in its *saturation region*.

As $V_{CE}$ rises the collector current rises but then levels out. In this levelling out region the BJT is said to be in its *active region*. The voltage at which the current levels out is referred to as $V_{CE(SAT)}$ and is typically in the range 0.2–1.0 V. When the BJT is in its active region we have (ignoring leakage currents),

$$I_C = \beta I_B$$

Note that the characteristic is not quite horizontal but does rise slightly with

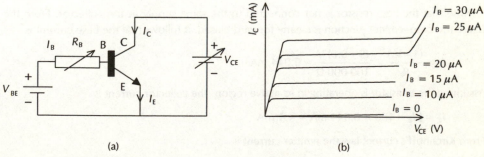

(a)                    (b)

*Figure 15.5*  **The CE configuration and output characteristics**

increasing $V_{CE}$. This is only a small effect and is usually ignored when determining circuit performance.

In the active region note that the collector current is almost independent of $V_{CE}$, being almost completely determined by the base current. Note also that as the base current increases so the collector current increases and that the characteristics are roughly evenly spaced.

Finally, beyond a certain value of $V_{CE}$, there is a rapid rise in the collector current. This can be traced back to the fact that the base-collector region is being operated as a reverse biased diode and therefore we identify this large increase in current as due to the phenomenon of breakdown discussed previously for reverse biased diodes. In practice, the manufacturer of the device will quote a $V_{CE(MAX)}$ which must not be exceeded.

━━━ **WORKED EXAMPLE 15.1**

---

Analyze the circuit shown in Figure 15.6 and obtain the base and emitter currents. Hence determine all p.d.s. Take $\beta = 100$ and when operating in its active region $V_{BE} = 0.6$ V.

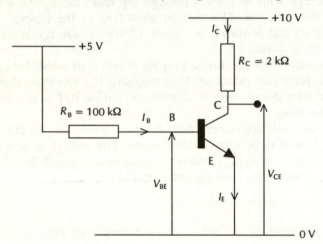

*Figure 15.6*

– Note that the base resistor is not connected to the same supply as the collector. From the figure the base–emitter junction is clearly forward biased. It follows that the base current is

$$I_B = \frac{(5 - V_{BE})}{R_B} = \frac{(5 - 0.6)\ \text{V}}{100\ 000\ \Omega} = 0.044\ \text{mA}$$

Assuming the transistor is operating in its active region, the collector current is

$$I_C = \beta I_B = 100 \times 0.044\ \text{mA} = 4.4\ \text{mA}$$

From Kirchhoff's current law the emitter current is

$$I_E = I_C + I_B \approx I_C = 4.4\ \text{mA}$$

The p.d. across the load resistor is

$$I_C \times R_C = (4.4 \times 10^{-3}\,\text{A}) \times (2000\,\Omega) = 8.8\,\text{V}$$

and it follows that the p.d. between the collector–emitter junction is

$$V_{CE} = (10 - 8.8)\,\text{V} = 1.2\,\text{V}$$

In the following we first describe the use of the BJT as a switch and then as an amplifier.

## 15.2   The BJT as a switch

The use of a BJT as a switch is straightforward. The object is to make use of a low base current to control the flow of the much larger collector current. In this manner the BJT can be thought of as a switch. The circuit shown in Figure 15.7 shows a typical arrangement using a BJT as a switch.

When the switch, S, is at position 1 the base and the emitter are fixed at the same potential. If they are at the same potential, the potential *difference* between the base and emitter must be zero, i.e. $V_{BE} = 0$ V. But, the base and emitter form a PN junction (see Figures 15.1 and 15.2), so that if $V_{BE}$ is equal to zero the junction is reverse biased. Recall that for a PN junction to be forward biased we not only require the p.d. between the P- and N-type materials to be connected in the correct manner (i.e. of the correct polarity), but we also need the p.d. between the base and emitter to be larger than the built-in or contact potential ($\phi \approx 0.6$–$0.7$ V for silicon and $\approx 0.2$–$0.3$ V for germanium) before the diode will conduct (see Section 14.3).

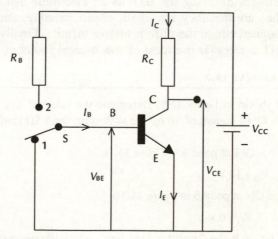

*Figure 15.7*   **The BJT as a switch**

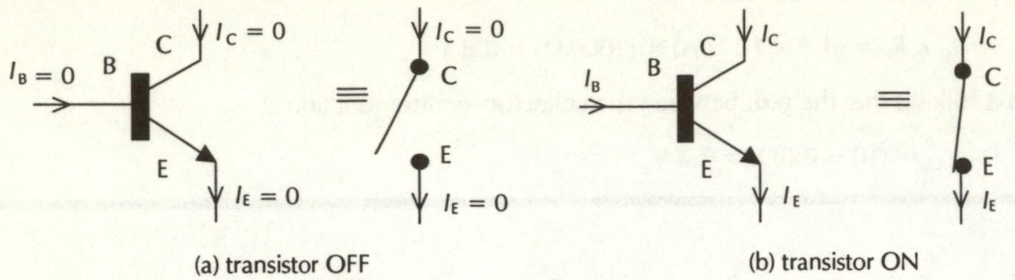

(a) transistor OFF                                    (b) transistor ON

*Figure 15.8*  **The BJT used as a switch**

Thus, the base emitter forms a reverse biased PN junction and, therefore, there is no current across this junction (as usual we ignore the leakage current). It follows that the collector current, $I_C$, is also zero. If no current flows through the load resistor $R_C$ it follows that the collector–emitter p.d. is equal to the DC supply voltage, $V_{CC}$. The transistor is described as being OFF and behaves as a very high resistance.

Now let us consider how the circuit operates when the switch is in position 2. In this case there is a current, $I_B$, through the base. The base and the emitter are no longer fixed at the same potential and the base–emitter PN junction is forward biased. According to Figure 15.4(b), which shows the common–emitter input characteristics, when a base current flows the p.d. across the base–emitter junction, $V_{BE}$, is approximately fixed at 0.6 V and independent of the p.d. across the collector–emitter junction, $V_{CE}$. Therefore the base–emitter junction is a forward biased PN junction and a current can flow through the load resistor $R_C$. Since there is a current through the transistor, it is described as being ON, and behaves as a low-resistance circuit element. Most of the p.d. is developed across the load resistor and, therefore, $I_C \times R_C \approx V_{CC}$, the DC supply voltage, so that $V_{CE} \approx 0$ V. These results are summarized in Figure 15.8.

The main advantages of using the BJT as an electronic switch compared with mechanical switches are that they are small, cheap, reliable, and have no moving parts. In a well designed circuit their life is almost infinite. Finally, they can be used to switch on and off at rates far in excess of mechanical switches.

■■■■ **WORKED EXAMPLE 15.2**

Consider the circuit shown in Figure 15.9. Determine the value of the resistor $R_B$ such that when the transistor is ON a current of 10 mA flows through the 1 kΩ load resistor. Assume the transistor has $\beta = 100$.

– When the transistor is OFF at point A in Figure 15.10

$$I_C = 0 \text{ A}, \qquad V_{CE} \approx 10 \text{ V}$$

When the transistor is ON at point B in Figure 15.10

$$I_C = 10 \text{ mA}, \qquad V_{CE} \approx 0 \text{ V}$$

Also shown in the figure is the 'transistor load line', which shows that the device can be operated anywhere along the line by the use of the appropriate base current. This will be discussed later on.

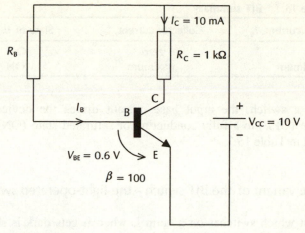

*Figure 15.9* **The transistor is ON**

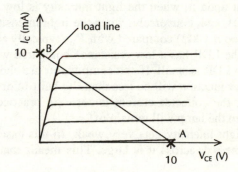

*Figure 15.10* **The transistor output characteristics with the load line**

Because we are using it as a switch, when OFF we have zero base current, and when ON the base current must have that value which will turn the transistor fully on in order for a collector current of 10 mA to flow. We are given $\beta = 100$, so from (15.1) the appropriate base current must be

$$I_B = \frac{I_C}{\beta} = \frac{10 \text{ mA}}{100} = 100 \text{ μA}$$

Since the transistor is ON we have

$$V_{BE} \approx 0.6 \text{ V}$$

so that the p.d. across the resistor $R_B$ is

$$10 \text{ V} - 0.6 \text{ V} = 9.4 \text{ V}$$

It follows that $R_B$ has the value

$$R_B = 94 \text{ k}\Omega$$

*Table 15.1* **BJT summary**

| Base current, $I_B$ | Collector current, $I_C$ | State of device |
|---|---|---|
| zero | zero | OFF |
| maximum | maximum | ON |

When used as a switch, the input base current drives the device from a non-conducting state (OFF) to a fully conducting or saturated state (ON). These results are summarized in Table 15.1.

## A simple variant of the BJT switch – the light-operated switch

A simple circuit which switches on a lamp L when it gets dark is shown in Figure 15.11. The resistor, $R_B$, and the light-dependent resistor, LDR, form a potential divider across the supply. The LDR is a resistor whose resistance depends upon the light intensity incident upon it: when the light intensity is low its resistance is high (typically about 1 MΩ), and, conversely, when the light intensity is high it has a low resistance (typically about 1 kΩ) compared with $R_B$ (typically about 10 kΩ).

When it is bright, the LDR has a low resistance compared with $R_B$ and it follows that the p.d. across the LDR is small. If the components are chosen such that the p.d. across the base–emitter junction will be less than the built-in or contact potential and the transistor is OFF, the collector current is zero (in practice very small). If the collector current is zero the lamp will remain off.

Suppose now the light intensity gets very weak. In this case the LDR has a very large resistance and the p.d. across it is large. This means that the transistor is ON

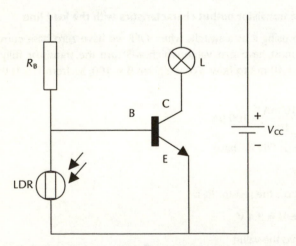

*Figure 15.11* **A simple light-operated switch. Note the symbols for the LDR and the lamp L**

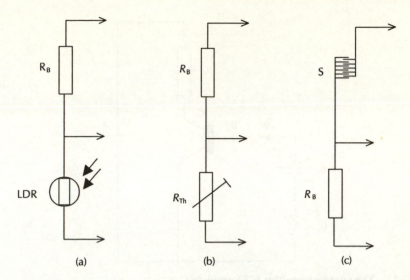

*Figure 15.12* **The uses of other resistive transducers in simple switch circuits. (a) Light-operated switch, (b) heat-operated switch, (c) moisture-operated switch**

and the transistor conducts. If the transistor allows a current to flow then the lamp will light.

The LDR is also known as a *resistive transducer* since its resistance changes with the amount of light incident upon it. By the appropriate change of resistive transducer, we can imagine arranging that the switch is designed to act as, for example, a heat or moisture detector; see Figure 15.12.

In Figure 15.12(a) we have the previously described LDR circuit. Figure 15.12(b) shows how we would arrange the circuit to act as a heat-operated switch by replacing the LDR by the thermistor. The thermistor is a resistor made of a semiconducting material having a resistance that decreases as its temperature rises (see Sections 5.2 and 14.1). The heat-operated switch operates in much the same manner as the LDR circuit. Figure 15.12(c) shows a moisture-operated switch circuit. The sensor S consists of two sets of conductors on a strip of board. The operation of this switch depends on the fact that if the gaps separating the conductors are dry the resistance of the sensor is high, while if wet the resistance of the sensor is small.

## 15.3 The BJT as an amplifier

We now discuss how the BJT is used to amplify a signal. We present a simple example of one of the most commonly encountered amplifier configurations, i.e. the common emitter amplifier.

Consider the circuit shown in Figure 15.13, where the transistor is connected in series with a DC supply $V_{CC}$ and a collector resistor $R_C$. Also, connected across the

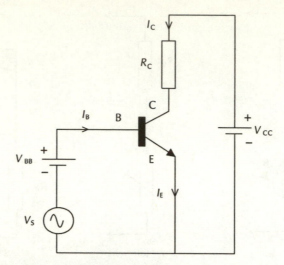

*Figure 15.13* **The common-emitter (CE) amplifier**

base–emitter junction is a DC supply $V_{BB}$ and a time-dependent signal given as $V_S(t)$. Suppose that the signal takes the form of a sinusoidal wave:

$$V_S(t) = V_p \sin(\omega t)$$

then the voltage wave applied between the base and the emitter, $V_{BE}$, consists of the sinusoidal signal $V_S(t)$ superimposed on the DC supply $V_{BB}$.

Consider the input characteristic. Since the p.d. across the base–emitter junction is varying we can see from Figure 15.13 that the current through the base, $I_B$, will vary in time about a mean value, $I_{BQ}$, and we can therefore write

$$I_B(t) = I_{BQ} + i_B(t)$$

Assuming that the DC supply $V_{BB}$ has been chosen such that the base-emitter is being operated in the linear region and that the amplitude of the signal does not cause the total p.d. across the base–emitter junction to fall below the linear region, then the current component $i_B(t)$ will also be sinusoidal with the same frequency and in phase with the source. This is indicated in Figure 15.14.

Technically, we say that the input signal which is to be amplified, $V_S$, is *biased* by the supply $V_{BB}$. This is necessary to achieve a linear wave form of the base current. The DC value $V_{BB}$ is referred to as the *quiescent* or the *operating* value. Essentially, although the input signal may go through positive and negative values, the effect of biasing is to ensure that the signal $V_S$ merely goes through values which are positive or negative with respect to the supply $V_{BB}$.

Let us now consider the effect of this varying base current on the collector current and on the p.d. across the load resistor $R_C$. For this we need to consider the output (collector) characteristics, i.e. $I_C$ versus $V_{CE}$, recall Figure 15.5(b), which we

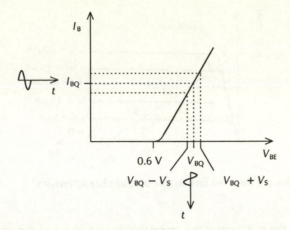

*Figure 15.14* **Operating the base–emitter junction in the linear region**

reproduce here as Figure 15.15. Note that we have also included the so-called *load line*. This will now be explained.

As we vary the base current, $I_B$, we have noted that the collector current will also vary (see Figure 15.15). If the collector current varies this means that the p.d. across the load resistor, $R_C$, will also vary (this is just Ohm's law; see Chapter 5). What happens to the p.d. between the collector and the emitter, $V_{CE}$? We can determine $V_{CE}$ by noting that at all times the p.d. across the transistor and the load resistor must add up to the supply voltage, i.e.

$$V_{CC} = V_{CE} + I_C R_C$$

This is merely a very simple application of Kirchhoff's voltage law applied to the circuit. This tells us that as we vary the collector current, the p.d. across the load resistor changes and the p.d. across the transistor, $V_{CE}$, must also change so that at all times they add up to the supply voltage. We can rewrite the above equation in the form

$$I_C = -\frac{1}{R_C} V_{CE} + \frac{V_{CC}}{R_C}$$

so that if we plot $I_C$ versus $V_{CE}$ the resulting graph is a straight line with gradient $-1/R_C$ and intercept on the current axis of $V_{CC}/R_C$ (see the Appendix). This line is referred to as the load line. Figure 15.15 is a plot of the load line superimposed on the output characteristics. This means that as we change the base current we can determine not only what the current through the collector is, but also the p.d. between the collector and emitter. Note that Worked Examples 15.1 and 15.2 effectively made use of the load line concept.    It follows that when there is no current through the load resistor (therefore the p.d. across the load resistor is zero

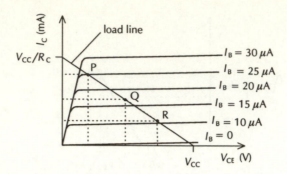

*Figure 15.15* **Introducing the load line to the output characteristics**

since $I_C \times R_C = 0$), the p.d. across the transistor is just $V_{CC}$ and the maximum current through the load resistor occurs when $V_{CE}$ is zero, i.e.

$$I_{C(max)} = \frac{V_{CC}}{R_C}$$

The load line is simply the plot of the collector–emitter p.d., $V_{CE}$, against the collector current, $I_C$ and is the locus of the only pairs of $I_C$ and $V_{CE}$ that are allowed for given values of the DC supply voltage, $V_{CC}$, and the load resistor $R_C$.

Also plotted in Figure 15.15 are the points P, Q and R. Note that as the base current varies the collector current varies in the same sense and with the same frequency. Since the collector current through the load resistor changes, the p.d. across the load resistor changes. It follows that the p.d. across the resistor, $V_{CE}$, also changes with the same frequency as the input current, but in opposite phase. This follows since as the collector current rises, the p.d. across the load resistor increases and, therefore, the p.d. across the transistor must fall.

In this manner we see that the transistor is being used as a voltage amplifier. By selecting a suitable DC input current to the base the transistor is turned partly ON. Any input AC signal will vary this input current and cause the collector current to vary. When the collector current flows through a resistor there is a p.d. developed across it. This implies that the p.d. across the transistor will vary.

## 15.4 The junction gate field effect transistor (JFET)

There are basically two types of FETs: the junction gate FET (JFET) and the metal oxide silicon FET (MOSFET). The difference between these FETs is in their construction. In the following we discuss only the JFET, in particular the N-channel JFET.

The operation of the N-channel junction gate field effect transistor (JFET) can be visualized by considering Figure 15.16, which depicts a simplified model of the

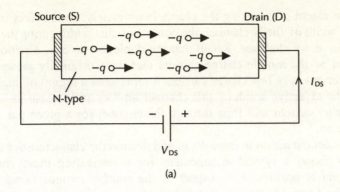

(a)

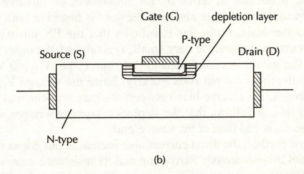

(b)

*Figure 15.16* **(a) Current through a silicon channel. (b) The junction gate field effect transistor (JFET)**

device. Although this model is not used in practical devices it provides an adequate model from which we may discuss how the device operates.

Figure 15.16(a) shows a silicon bar which is called the channel (N-channel if made of N-type material, and P-channel if made of P-type material; see Section 14.2). The contacts made on the ends of the two faces of the bar are called the *source* (S) and *drain* (D), respectively. With the battery as connected in the figure, electrons are injected into the channel at the source and collected at the drain. The current, $I_{DS}$, flows from drain to source as shown and the magnitude of the current depends on the resistance offered to the flow of current by the channel.

The action of the JFET depends on providing a mechanism for varying the resistance of the channel and, therefore, the current $I_{DS}$. This is achieved by introducing a P-type region into the silicon bar. A PN junction is therefore formed (see Section 14.3). This PN junction is known as a *gate* (G); see Figure 15.16(b).

Note that with no connections made to any terminal the depletion layer associated with the PN junction extends partly into the N-type channel between the source and drain (also into the P-type gate region). Usually the N-channel is lightly doped compared with the P-type gate and the depletion layer extends well into the channel but not to the extent of blocking the channel completely. This effect is to decrease the *effective* width of the N-channel; the control of this width forms the whole basis of the device, as we shall see. Recall that the depletion layer is a region which has been depleted of the mobile charge carriers that were originally present. Since the depletion layer has very few charge carriers it represents a region of high resistance. Decreasing the effective width of this channel implies an increase in its resistance, so that a smaller current will flow through the channel for a given p.d. between the source and drain.

Let us consider the action in more detail and discuss the characteristics of the JFET. Figure 15.17 shows a typical arrangement for investigating these characteristics. Since the drain is positive with respect to the source, current flows through the channel in the form of electron conduction from source to drain (though, since conventional current is defined in terms of the movement of positive charge, the current is from drain to source). Note also that the gate is negative both with respect to the source and to the drain. From this it follows that the PN junction is reverse biased and the gate current is therefore very small, typically of the order of pA.

Let us first take the gate to be connected to the source, i.e. $V_{GS} = 0$ V. As $V_{DS}$ is increased from zero the drain current rises linearly. Since the voltage $V_{DS}$ is dropped along the channel length, the reverse bias between the gate and channel is greater at the drain end. From this, it follows that the depletion layer is wedged shaped, with the layer wider at the drain end than at the source end.

As $V_{DS}$ is increased further, the drain current also increases but not as rapidly. This is because the channel is continuously narrowing and its resistance increasing. This is referred to as the *Ohmic region* of behaviour; see Figure 15.18. In this region the JFET is acting as a variable-voltage resistor.

As $V_{DS}$ is increased even further, the above arguments might imply that eventually a point is reached when the depletion layer widens sufficiently to block the whole

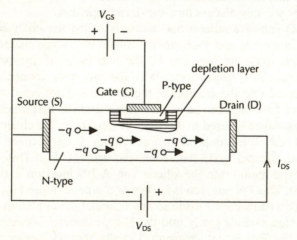

*Figure 15.17* **The principle of operation of the JFET**

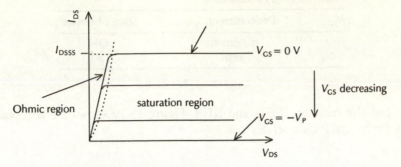

*Figure 15.18* **N-channel JFET characteristics**

channel and inhibit completely the flow of electrons. In fact this is not so, and the channel reduces to an extremely narrow filament and the electron flow is limited to a virtually constant rate, independent of the value of $V_{DS}$. The channel is described as *saturated* and the drain current is referred to as the *drain–source saturation current* ($I_{DSSS}$), see Figure 15.18.

As $V_{GS}$ is increased, the gate becomes more negative with respect to both the source and the drain, so that the PN junction becomes more reversed biased. Recalling previous work (see Section 14.3), this implies that the width of the depletion layer increases (see Figure 14.12). The depletion layer, therefore, extends further into the N-channel, thus reducing the effective width for the electrons to flow from source to drain (recall that the depletion layer is a region of high resistance since it has very few mobile charge carriers). This, in effect, increases the resistance offered to the flow of electrons from source to drain, and so the drain current becomes smaller. As $V_{GS}$ is increased further, a point is reached when the depletion layer extends completely across the channel and no current can flow (except for the small leakage current of a reverse biased PN junction; see Section 14.3). At this critical voltage, the channel is said to be *pinched off* and the critical voltage is referred to as the *gate pinch-off voltage, $V_P$.*

Thus, control of the current through the JFET occurs by narrowing the channel so that it becomes depleted of charge carriers. For this reason, JFETs are said to operate in the *depletion mode*. These results are summarized in Table 15.2, where $V_P$ is the pinch-off voltage, as discussed above. Compare these results with the effect of the base current on the collector current for a BJT shown in Table 15.1 (see Sections 15.1 and 15.2).

Notice that it is the input voltage between the gate and the source, $V_{GS}$, which controls the drain current ($I_{DS}$). It follows that the JFET is a voltage-controlled device.

### The JFET symbol

The symbol for the JFET must not only distinguish between N- and P-channel

*Table 15.2*   **JFET summary**

| $V_{GS}$ | Drain current, $I_{DS}$ | State of device |
| --- | --- | --- |
| 0 | maximum | ON |
| $-V_P$ | zero | OFF |

devices, but also between JFETs and BJTs. Figure 15.19 shows the accepted JFET symbols for N- and P-channel devices.

### ■■■ COMMENTS

- The names of the terminals come from the parts they play in the operation of the device:
  - The source is the *source* of electrons;
  - the drain *collects* the electrons;
  - the gate *controls* the flow of the electrons from the source to the drain.

  Thus, the terms given to each of the terminals accurately describe their functions.
- The JFET configuration described above is termed a common source. This is obvious notation since the source is common to both the drain and the gate.

## 15.5   The JFET as a switch

The use of a JFET as a switch is straightforward. The circuit shown in Figure 15.20 shows a typical arrangement using a JFET as a switch. The object is to make use of a p.d. between the gate and source, $V_{GS}$, to control the drain current, $I_{DS}$, and therefore, the p.d. between the drain and source, $V_{out}$ in Figure 15.20. In this manner the JFET can be thought of as a switch.

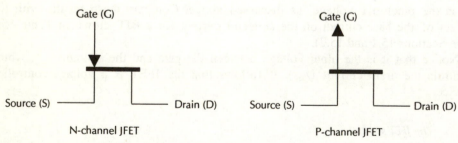

*Figure 15.19*   **Symbols for the JFET**

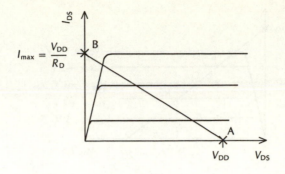

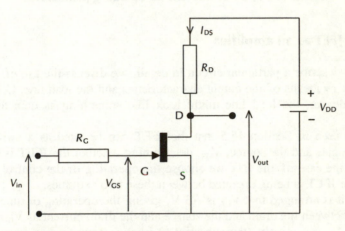

*Figure 15.20* **The JFET switch output characteristics and circuit**

Also shown in the figure for the output characteristics is the load line. In a similar fashion to the discussion for the BJT, the p.d. across the JFET and the load resistor, $R_D$, must add up at all times to the supply voltage, i.e.

$$V_{DD} = V_{DS} + I_{DS}R_D$$

and, again, is merely a very simple application of Kirchhoff's voltage law applied to the circuit.

At point A, when $I_{DS}$ is equal to zero, $V_{DS}$ is approximately equal to the supply voltage, $V_{DD}$, while at point B, when the drain current is a maximum, since $V_{GS}$ is equal to zero, $V_{DS}$ is equal to zero. The maximum current flowing through the transistor is

$$I_{max} = \frac{V_{DD}}{R_D}$$

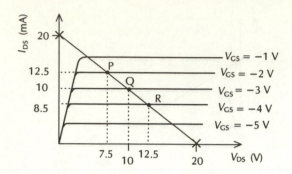

*Figure 15.21*   **The load line of a JFET and how the operating point moves**

### 15.6   The JFET as an amplifier

Rather than describe a particular circuit in detail, we discuss the use of the JFET as an amplifier by means of the output characteristics and the load line. Considering an N-channel device, the load line might look like something as depicted in Figure 15.21.

We have seen in Section 15.5 that the JFET can be used as a switch, the p.d. between the gate and the source, $V_{GS}$, determining whether the JFET is ON or OFF. Just like in the case of the BJT we choose the operating or the control signal to be such that the JFET is being operated between these two extremes.

Suppose it is arranged that $V_{GS}$ is $-3$ V, giving the operating or quiescent values of the p.d. between the drain and the source and the drain current as $V_{DSQ} = 10$ V and $I_{DSQ} = 10$ mA, respectively (the subscript Q implies operating or quiescent value); see Figure 15.21. Any input signal connected to the amplifier would cause the operating point to move just as for the BJT (see Section 15.3). For example, if $V_{GS}$ changes from $-2$ V to $-4$ V, the drain current changes from 12.5 mA to 8.5 mA, while $V_{DS}$ changes from 7.5 V to 12.5 V (these figures are estimated from the output characteristics). Thus, a change in the input signal of 2 V causes the output to change by 12.5 V − 7.5 V = 5 V.

### 15.7   Practical aspects of transistors

There are several thousand types of transistors. Each transistor is given a code by the manufacturer that identifies it. American manufacturers use a code that begins with the letters '2N' followed by a number, for example 2N3705 (diodes have a code that begins with '1N'). European manufacturers use a code which indicates the semiconductor material, 'A' for germanium and 'B' for silicon, and the intended main use of the transistor, 'C' indicates it can be used as an audio frequency amplifier, 'F' indicates it can be used for radio frequency amplification, and 'S' indicates its main use would be for switching. Thus, a transistor marked as BC108 implies it is made out of silicon and it is to be used for audio frequency amplification. Note that some manufacturers have their own codes.

While one type of transistor may replace another in many circuits, it is often useful to study the published data supplied by the manufacturer when making a choice. Table 15.3 lists the main properties or transistor parameters for several of the more commonly used BJTs. Similar tables can be found for FETs.

### Current, voltage and power ratings

In most cases the symbols used are self-explanatory:

- $I_C$ max is the maximum collector current the transistor can pass without damage.
- $V_{CEO}$ max is the maximum p.d. that can be applied between the collector and emitter when the base is open circuited.
- $V_{EBO}$ max is the maximum p.d. that can be applied between the emitter and the base when the collector is open circuited.
- $P_{tot}$ is the maximum total power rating at 25°C air temperature and is approximately equal to $V_{CE} \times I_C$.

### DC current gain (β)

Due to manufacturer spreads the DC current gain, $\beta$, is not the same for all transistors of the same type. Usually minimum and maximum values are quoted, or sometimes just the minimum or the typical value. Also, since $\beta$ decreases at high and low collector currents, the current at which it is measured is stated. It also depends upon temperature. When selecting a transistor type for a particular circuit, we have to ensure that its minimum $\beta$ will give the current gain required by that circuit – the fact that $\beta$ may be larger is not usually of great importance. In many applications $\beta$ is the most important factor – this is true for small-signal audio frequency voltage amplifiers since the current, voltage and power ratings are usually within the acceptable bounds for most transistors.

Table 15.3

|  | BC108 | BF194 | 2N3053 | 2N3705 |
|---|---|---|---|---|
| $I_C$ max (mA) | 100 | 30 | 1000 | 500 |
| β | 110–800 | 115 (typical) | 50–250 | 50–150 |
| at $I_C$ mA | 2 | 1 | 150 | 50 |
| $P_{tot}$ (mW) | 360 | 220 | 800 | 360 |
| $V_{CEO}$ max (V) | 20 | 20 | 40 | 30 |
| $V_{EBO}$ max (V) | 5 | 5 | 5 | 5 |
| $f_T$ (MHz) | 250 | 260 | 100 | 100 |
| Outline | TO18 | MM10b | TO39 | TO92a |

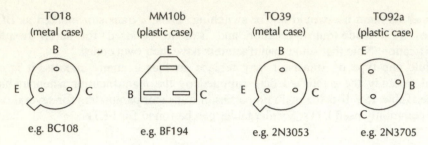

*Figure 15.22*   **Transistor outlines**

### Transition frequency, $f_T$

This is the frequency at which $\beta$ falls to unity – this is important in high-frequency circuits.

### Outlines

The case which actually protects the actual transistor is known as the *encapsulation*. The two most commonly used encapsulations are metal and plastics. The shape, or outline, of the transistor is chosen by the manufacturer and depends upon what it is used for. The outlines for the transistors given in Table 15.3 are shown in Figure 15.22.

### 15.8  Summary

We presented a description of the BJT, restricting our discussion to the NPN device, noting that the description of the PNP device was straightforward to derive. We began by considering the charge flow and current in the device and then on to a description of the current voltage characteristics. These characteristics led us on to describe the use of the BJT as a switch. We then described the use of the BJT as an amplifier. We note that the BJT is a current-controlled device, since changes in the base current determine its operation.

We then presented a description of the FET, restricting our discussion to the N channel JFET device, noting that the description of the P channel device is derived by replacing N-type material by P-type, and vice versa, and electrons by holes. We began by considering the charge flow and current in the device and then described the JFET voltage characteristics. These characteristics led us to describe the FET as a voltage-controlled device. We then followed this with descriptions of the uses of the device as a switch and as an amplifier. For the amplifier, particular emphasis was placed on the output characteristics and the load line to indicate the action of amplification.

We finished our introduction to the transistor with a discussion of some practical aspects of transistors. This discussion was restricted to the BJT, with the comment that similar notes could be made for the FET.

## ▬▬ Problems

**15.1**  Analyze the circuit shown in Figure P15.1 and obtain the base and emitter currents. Hence determine all p.d.s. Take $\beta = 150$ and when operating in its active region, $V_{BE} = 0.6$ V.

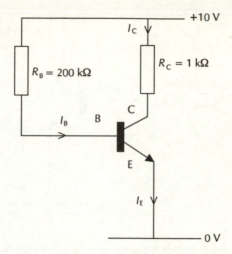

**Figure P15.1**

>  *Hint:* Assume that the transistor is operating in its active region.

**15.2**  Analyze the circuit shown in Figure P15.2 and obtain the base and emitter currents. Hence determine all p.d.s. Take $\beta = 250$ and when operating in its active region, $V_{BE} = 0.6$ V.

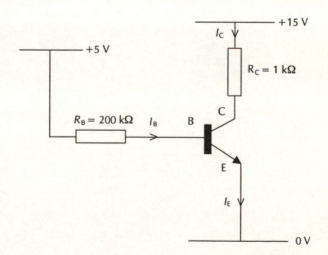

**Figure P15.2**

>  *Hint:* Assume that the transistor is operating in its active region.

**15.3** Analyze the circuit shown in Figure P15.3 and obtain the base and emitter currents. Hence determine all p.d.s. Take $\beta = 100$ and when operating in its active region, $V_{BE} = 0.6$ V. Comment on your results.

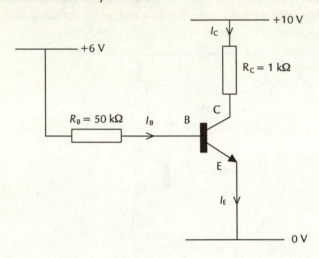

*Figure P15.3*

*Hint:* Assume that the transistor is operating in its active region.

# *Appendix*

Mathematics may be thought of as a language that is very useful in science – in general and engineering in particular – and the use of mathematics, as such, is a skill in which engineers must become proficient. In this Appendix we present a brief review of the operations and methods that are useful for our needs.

## A.1 Algebra

We now discuss some basic algebraic manipulations of equations. It is particularly useful to be able to change around an equation. Suppose $x$, $y$ and $z$ are three variables such that

$$x = \frac{y}{z} \tag{A.1}$$

In this form, $x$ is said to be the subject of the equation and we can determine its value once we are given $y$ and $z$. The term 'subject of the equation' merely means that we can write '$x$ = something'.

Case 1    Suppose, however, that $x$ and $z$ are known and $y$ is unknown. We then have to manipulate (A.1) so that $y$ is the subject. If we multiply both sides of (A.1) by $z$ (remember, $z$ is just some given number), then the equation will still be true. We have

$$x \times z = \frac{y}{z} \times z$$

The $z$s on the right-hand side cancel, so that we have

$$x \times z = \frac{y}{\not z} = \frac{y}{\not z} = y$$

which we write as

$$y = x \times z$$

*Case 2*  Next, suppose that $x$ and $y$ are the known quantities and that $z$ is now unknown. We now have to manipulate (A.1) so that $z$ is the subject. If we proceed as previously and multiply both sides of (A.1) by $z$ we find

$$y = x \times z$$

This operation has the desirable result in bringing the unknown $z$ onto the top, but we have not quite reached our final result '$z =$ something'. For this, we have to get rid of the $x$ variable. This is straightforward, since dividing both sides by the variable $x$ gives

$$\frac{y}{x} = \frac{\not{x} \times z}{\not{x}} = z$$

Again, it is customary to write our result in the form 'subject = something'

$$z = \frac{y}{x}$$

The next type of equation we wish to discuss takes the form

$$y = mx + c \tag{A.2}$$

where $m$ and $c$ are given numbers. Here we have written y as the subject. Another common way of expressing this relationship is to say that $y$ is a function of $x$, so that given a value of $x$ we can determine the corresponding value of $y$.

We will discuss this equation in more detail in Section A.2. For the moment we merely wish to arrange the equation so that $x$ is the subject, i.e. we wish to determine the value that $x$ must take so that $y$ has a particular value. We indicate the steps to achieve this. If we subtract $c$ from both sides of the equation we have

$$y - c = mx + \not{c} - \not{c}$$

On the right-hand side the $c$s cancel out so that we have

$$y - c = mx$$

Now divide both sides by $m$:

$$\frac{y - c}{m} = \frac{\not{m}x}{\not{m}} = x$$

The final result is

$$x = \frac{y - c}{m}$$

Notice the general rule for manipulating an equation: what we do to one side of the equation, we also do to the other side.

The following rules of multiplying, dividing, adding and subtracting fractions are useful, where $a$, $b$ and $c$ are three numbers:

Multiplying: $\left(\dfrac{a}{c}\right) \times \left(\dfrac{b}{d}\right) = \left(\dfrac{a \times b}{c \times d}\right)$ 　　　　　　　　(A.3)

Dividing: $\dfrac{a/b}{c/d} = \left(\dfrac{a \times d}{b \times c}\right)$ 　　　　　　　　(A.4)

Adding: $\dfrac{a}{c} \pm \dfrac{b}{d} = \dfrac{a \times d \pm b \times c}{c \times d}$ 　　　　　　　　(A.5)

**████ WORKED EXAMPLE A.1**

In the following equation a is some given number. Solve the equation for the unknown x.

$$a = \frac{1}{1+x}$$

− In the equation, not only does x not appear on top, it also does not occur by itself. We first arrange the equation so that $1 + x$ appears on top. We achieve this by multiplying both sides by $1 + x$. We find

$$a \times (1 + x) = 1$$

Our next step is to try and arrange the equation so that x appears by itself. We cannot achieve this in one step. Rather we first divide both sides by a. This gives the result

$$1 + x = \frac{1}{a}$$

from which it is straightforward to show

$$x = \frac{1}{a} - 1 = \frac{1-a}{a}$$

where we have made use of (A.5).

## A.2  Equations

The equation

$$y = mx + c \tag{A.2}$$

is of great importance in engineering – an enormous number of relationships between variables can be written in this form. A useful way of visualizing the relation between the two quantities $x$ and $y$ is by a graph which is a plot of corresponding pairs of $x, y$ values. The result of plotting $y$ versus $x$ is shown in Figure A.1.

The equation (A.2) is referred to as linear because the graph of $y$ versus $x$ is a straight line. Note the following:

- Each point on the line is made up of an $x$ value and a $y$ value, where the $y$ value is related to the $x$ value by (A.2). The $x$ and $y$ values are called the coordinates of the point and are recorded as $(x, y)$.
- The constant $c$ is called the intercept and represents the value of $y$ at which the straight line intercepts the $y$ axis. Thus, the coordinates of the intercept on the $y$ axis are $(0, c)$.
- The constant $m$ is called the gradient or slope of the line, and is equal to the tangent of the angle that the line makes with the $x$ axis

$$m = \tan(\theta)$$

- We need only two points to determine a straight line. If $(x_1, y_1)$ and $(x_2, y_2)$ are any two points on the straight line then the gradient of the line can be expressed as

$$\text{gradient} = \frac{y_2 - y_1}{x_2 - x_1} = \frac{\Delta y}{\Delta x}$$

where $\Delta y$ is the *change* in the $y$ values, and $\Delta x$ is the corresponding *change* in the $x$ values between the two points.

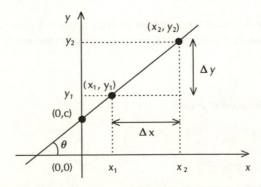

*Figure A.1*  **The straight line graph**

■ The values of $m$ and $c$ can be either positive or negative. If $m$ is positive the line has a positive gradient, which implies that $y$ increases as $x$ increases. Conversely, if $m$ is negative the line has a negative gradient, which implies that $y$ decreases as $x$ increases.

■■■■ **WORKED EXAMPLE A.2**

Plot the following equations

(i) $y = -2x + 4$,   (ii) $y = 3x - 2$

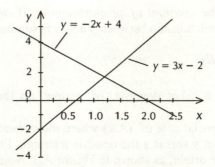

*Figure A.2* **Plotting the lines $y = -2x + 4$ and $y = 3x - 2$**

## A.3 Proportionality

A very important mathematical operation in engineering is determining the relation between two sets of measurements. In the following we discuss two types of relationship between measurements:

*direct proportion* and *inverse proportion*

### Direct proportion

Let us suppose that we perform some experiment and two sets of readings are obtained for the quantities $x$ and $y$, as in Table A.1. We note that when $x$ is doubled, $y$ doubles; when $x$ is trebled, $y$ trebles; when $x$ is increased by a factor of four, $y$ is

Table A.1

| $x$ | 1 | 2 | 3 | 4 |
|-----|---|---|---|---|
| $y$ | 2 | 4 | 6 | 8 |

increased by a factor of four. There is a one-to-one correspondence between each value of the $(x, y)$ pair, and we say that $y$ *is* directly proportional to $x$ or that $y$ varies directly as $x$. We express this result mathematically by writing

$$y \propto x \tag{A.6}$$

Also, the ratio of corresponding $x$, $y$ pairs has a constant value. For the data presented in Table A.1 this value is 2, i.e.

$$\frac{y}{x} = \text{constant} = 2$$

The constant is called the *constant of proportionality*. If we give this constant a symbol, say $K$, then the relationship between $y$ and $x$ is summed up by the equation

$$\frac{y}{x} = K \quad \text{or} \quad y = Kx \tag{A.7}$$

The precise value that $K$ takes depends, of course. on the detailed relationship between $y$ and $x$.

Note that (A.7) is a special case of (A.2) where the constant $c$ is equal to zero. It follows that when we plot $y$ versus $x$ the result is a straight line graph of gradient $K$ which passes through the origin, as shown in Figure A.3. Note that as the constant $K$ increases in size, the resulting straight line makes a steeper angle with respect to the $x$ axis.

### Inverse proportion

One other case of interest will now be discussed. Let us consider the two sets of readings for the quantities $x$ and $y$ as in Table A.2. There is again a one-to-one

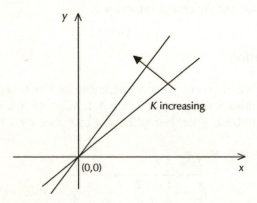

*Figure A.3*  **Direct proportion**

*Table A.2*

| x | 2 | 4 | 6 | 8 | 10 |
|---|---|---|---|---|----|
| y | 30 | 15 | 5 | 7.5 | 6 |

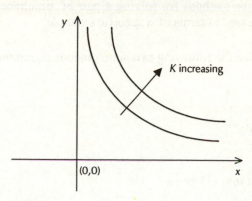

*Figure A.4* **Inverse proportion**

correspondence between corresponding values of $x$ and $y$. In this example, however, note that when $x$ is doubled, $y$ is halved; when $x$ is trebled, $y$ has one-third of its previous value. We say that $y$ is inversely proportional to $x$, or that $y$ varies inversely as $x$. This result is expressed mathematically by writing

$$y \propto \frac{1}{x} \tag{A.8}$$

Note that the product of corresponding $x$, $y$ pairs has a constant value and we write

$$y = \frac{K}{x} \quad \text{or} \quad xy = K \tag{A.9}$$

The precise value that $K$ takes depends, of course, on the detailed relationship between $y$ and $x$. For the data presented in Table A.2 the value of the constant is 60.

The result of plotting $y$ versus $x$ is shown in Figure A.4.

Note that if we plot $y$ versus $1/x$ then the resulting graph is a straight line passing through the origin.

## A.4 Simultaneous equations

Consider an equation such as $2x + 3y = 12$, which has two unknowns $x$ and $y$. Such an equation does not have a unique solution, i.e. it does not have a unique pair of values which satisfies this equation. For example $(x = 0, y = 4)$, $(x = 6, y = 0)$, $(x = 4.5, y = 1)$ all satisfy the equation and are therefore solutions.

If a problem has two unknowns, a unique solution is only possible if we have two equations. Similarly if we have a problem with 3, 4, 5, ... unknowns its solution requires 3, 4, 5, ... equations, respectively.

We now present three methods for solving a pair of simultaneous equations. The methods are best discussed in terms of a specific example.

*Method 1*   Solve the following two simultaneous equations

$$2x + 3y = 12 \tag{1}$$

$$x - y = 1 \tag{2}$$

From (2) we can write

$$y = x - 1$$

Substituting this result into (1) gives

$$2x + 3(x - 1) = 12$$
$$2x + 3x - 3 = 12$$
$$5x = 15$$

From this result it follows that

$$x = 3$$

We can substitute this result into either (1) or (2) to determine $y$. Hence, the desired solution is

$$(x, y) = (3, 2)$$

*Method 2*   Multiply each side of (2) by the factor 3, giving

$$3(x - y) = 3x - 3y = 3$$

Adding this result to (1) gives

$$\begin{array}{rl} 2x + 3y = & 12 \\ 3x - 3y = & 3 \\ \hline 5x \quad = & 15 \end{array}$$

Note that the terms in $y$ give zero, leading to an equation that just involves $x$. This is, of course, why we chose to multiply Equation (2) by the factor 3. Solving gives

$$x = 3$$

We can substitute this result into either (1) or (2). We find

$$y = 2$$

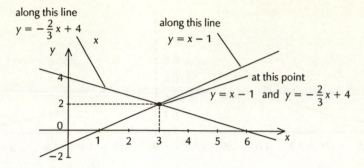

*Figure A.5* **Graphical solution**

Hence the solution is

$$(x, y) = (3, 2)$$

as determined previously.

*Method 3* We now demonstrate how two linear equations in two unknowns can also be solved by a graphical method. We begin by writing (1) and (2) in the form (A.2). We have

$$y = -\tfrac{2}{3}x + 4$$
$$y = x - 1$$

The result of plotting these equations is shown in Figure A.5. We note that the two lines intersect. The point of intersection represents the solution, since at this point *both* equation (1) and equation (2) are satisfied. From the graph the point of intersection is

$$(x, y) = (3, 2)$$

in agreement with the previous results.

## A.5 The exponential function and the natural logarithm

Quite often in engineering we find that a quantity $y$ is expressed as a power of some number $a$:

$$y = a^x$$

where $a$ is called the base number. In practice, the two bases most commonly used are base 10 and base e, where $e = 2.718\ldots$ is called the natural number. Here we shall only be concerned with the case

$$a = e$$

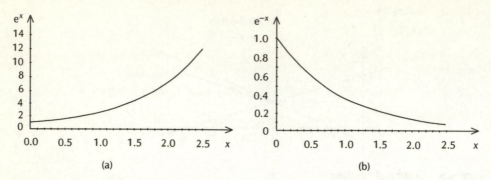

*Figure A.6*  **The variation of the functions (a) exp(x) and (b) exp(x) with x**

so that the relation between $y$ and $x$ is

$$y = \exp(x) \tag{A.10}$$

The graph of $\exp(x)$ versus $x$ is shown in Figure A.6(a). Note that the function has the value of unity at $x = 0$ and increases indefinitely as $x$ increases. Similarly, Figure A.6(b) shows how the function $\exp(x)$ varies with $x$. We note that the function has the value of unity at $x = 0$ but now decreases to zero as $x$ increases.

The exponential function (A.10) is well known and given $x$ it is straightforward to determine the corresponding value of $y$. Suppose, however, we are given the value of $y$ and we want to determine the value of $x$. How do we proceed? Put another way, what value of $x$ results in a given value for $y$? To solve this we introduce the natural logarithm, $\log_e$ or ln, with the property that

$$y = \exp(x) \Leftrightarrow x = \ln(y)$$

Just like the exponential function, the natural logarithm is well known; given a value of $y$, the corresponding value of $x$ is straightforward to determine.

■■■■ **WORKED EXAMPLE A.3**

Given that $y$ is related to $x$ according to (A.10), determine $x$ if $y = 48$.
– We have

$$48 = \exp(x)$$

so that $x$ is given as

$$x = \ln(48) = 3.871$$

As a check we have

$$\exp(3.871) = 48$$

Suppose we want to solve a more complicated relationship between $x$ and $y$, for example:

$$y = A(\exp(x) + b) \qquad (A.11)$$

where $A$ and $b$ are some given numbers. How do we proceed? We try to arrange the above so that we have

$$\exp(x) = \text{something}$$

since then we can write

$$x = \ln(\text{something})$$

It is straightforward to show that if $y$ is some given number, then the value of $x$ which satisfies (A.11) is

$$x = \ln\left(\frac{y}{A} - b\right)$$

## A.6 Angular measure

Consider a line which rotates about one of its end points. The 'amount of turn' the line goes through when rotated about one of its end points is measured by the angle $\theta$, see Figure A.7(a). If the line rotates through an angle such that it is in the same position it started from, the angle of rotation is customarily defined as $360°$, see Figure A.7(b). Mathematically, we prefer to measure angular rotation in the radian (rad). The radian is defined such that a rotation of $360°$ is equivalent in radian measure to an angle of $2\pi$, i.e.

$$360° = 2\pi \text{ rad}$$

From this we can determine that an angular rotation of 1 radian is equivalent to $57.296°$, since

$$1 \text{ rad} = \frac{360°}{2\pi} = 57.296°$$

Alternatively, a rotation through $1°$ is the same as a rotation through 0.0175 rad. This follows from the result

$$1° = \frac{2\pi}{360} = 0.0175 \text{ rad}$$

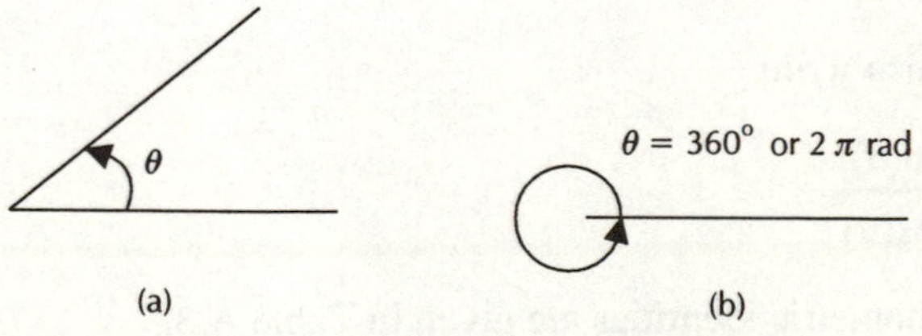

(a)                                    (b)

*Figure A.7* **The 'amount of turn' is measured by the angle θ**

Special cases

Although radian measure is the preferred way of discussing angles, by force of habit we tend to discuss angles in terms of degrees. It is useful to remember the following special cases:

$$\frac{\pi}{2} \text{ rad} = 90°$$

$$\frac{\pi}{3} \text{ rad} = 60°$$

$$\frac{\pi}{4} \text{ rad} = 45°$$

$$\frac{\pi}{6} \text{ rad} = 30°$$

Since $\pi \approx 3.142$ it is useful to remember that $1 \text{ rad} \approx 60°$ and $1° \approx 1/50 \text{ rad}$.

## A.7 Trigonometric functions

We begin by defining a right-angled triangle as a triangle which has one of its angles equal to $\pi/2$ (i.e. $90°$). Consider the right-angled triangle shown in Figure A.8, where the side $a$ is opposite the angle $\theta$, the side $b$ is adjacent to the angle $\theta$, and $c$ is the hypotenuse. The three basic trigonometric functions defined by a right-angled triangle are the sine (sin), cosine (cos), tangent (tan) functions. In terms of the angle $\theta$, these functions are defined by

$$\sin(\theta) = \frac{\text{side opposite } \theta}{\text{hypotenuse}} = \frac{a}{c}$$

$$\cos(\theta) = \frac{\text{side adjacent } \theta}{\text{hypotenuse}} = \frac{b}{c}$$

$$\tan(\theta) = \frac{\text{side opposite } \theta}{\text{side adjacent } \theta} = \frac{a}{b}$$

Note that we can also write

$$\tan(\theta) = \frac{\sin(\theta)}{\cos(\theta)}$$

Some useful trigonometric identities are given in Table A.3.
The variation of $\sin(\theta)$ and $\cos(\theta)$ for $0 \leqslant \theta \leqslant 2\pi$ is shown in Figure A.9.

*Table A.3*

$$\sin(\theta) = \cos(\pi/2 - \theta)$$
$$\cos(\theta) = \sin(\pi/2 - \theta)$$
$$\sin(-\theta) = -\sin(\theta)$$
$$\cos(-\theta) = \cos(\theta)$$
$$\sin^2(\theta) + \cos^2(\theta) = 1$$
$$\sin(2\theta) = 2\sin(\theta)\cos(\theta)$$
$$\cos(2\theta) = \cos^2(\theta) - \sin^2(\theta)$$
$$\sin(\theta + \phi) = \sin(\theta)\cos(\phi) + \sin(\phi)\cos(\theta)$$
$$\cos(\theta + \phi) = \cos(\theta)\cos(\phi) - \sin(\theta)\sin(\phi)$$

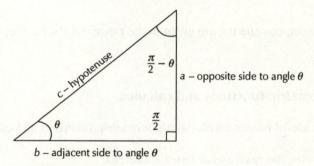

*Figure A.8* **The right-angled triangle**

From Figure A.8 we note in Table A.4 some special values of $\sin(\theta)$ and $\cos(\theta)$ which are useful to remember. We finally note that the sine, cosine and tangent functions are examples of functions which are termed periodic, i.e. they repeat themselves in a regular manner. The period of these functions is $2\pi$:

$$\sin(\theta + 2\pi) = \sin(\theta)$$
$$\cos(\theta + 2\pi) = \cos(\theta)$$
$$\tan(\theta + 2\pi) = \tan(\theta)$$

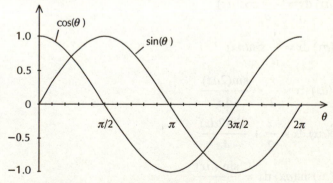

*Figure A.9* **The variation of the sine and cosine functions**

*Table A.4*

| θ(rad) | sin(θ) | cos(θ) |
|--------|--------|--------|
| 0 | 0 | 1 |
| $\pi/6$ | 0.5 | $\sqrt{3}/2 = 0.866$ |
| $\pi/3$ | $\sqrt{3}/2 = 0.866$ | 0.5 |
| $\pi/4$ | $1/\sqrt{2} = 0.707$ | $1/\sqrt{2} = 0.707$ |
| $\pi/2$ | 1 | 0 |
| $\pi$ | 0 | −1 |
| $3\pi/2$ | −1 | 0 |
| $2\pi$ | 1 | 0 |

Therefore, once sin, cos and tan are given in the range $0 \leqslant \theta \leqslant 2\pi$ they are known for all values of $\theta$.

## A.8  Trigonometric functions and calculus

We quote some useful results involving trigonometric functions and calculus:

*Trigonometric functions and differential calculus*

$$\frac{\mathrm{d}}{\mathrm{d}x} \sin(ax) = a\cos(ax)$$

$$\frac{\mathrm{d}}{\mathrm{d}x} \cos(ax) = -a\sin(ax)$$

*Trigonometric functions and integral calculus*   To the following indefinite integrals an arbitrary constant should be added

$$\int \sin(ax)\,\mathrm{d}x = -\frac{1}{a}\cos(ax)$$

$$\int \cos(ax)\,\mathrm{d}x = \frac{1}{a}\sin(ax)$$

$$\int \sin^2(ax)\,\mathrm{d}x = \frac{x}{2} - \frac{\sin(2ax)}{4a}$$

$$\int \cos^2(ax)\,\mathrm{d}x = \frac{x}{2} + \frac{\sin(2ax)}{4a}$$

$$\int \cos(ax)\sin(ax)\,\mathrm{d}x = \frac{\sin^2(ax)}{2a}$$

## Problems

**A.1**   Determine the value of $x$ if

(a)  $6 = \dfrac{15}{x}$,  (b)  $10 = \dfrac{x}{20}$,  (c)  $5 = \dfrac{1}{1-x}$,  (d)  $16 = \dfrac{10}{1+x}$

**A.2**   Determine the value of $x$ if

(a)  $2x + 3 = 7$,  (b)  $\dfrac{x}{3} + \dfrac{1}{4} = 0$,  (c)  $\dfrac{3}{x} + 2 = 5$

**A.3**   Confirm the following expressions

(a)  $\dfrac{1}{4} + \dfrac{2}{3} = \dfrac{11}{12}$  (b)  $\left(\dfrac{2}{3}\right) \times \left(\dfrac{3}{4}\right) = \dfrac{1}{2}$,  (c)  $\dfrac{(2/3)}{(3/4)} = \dfrac{8}{9}$

Verify your results by explicit calculation.

**A.4**   The equation of a straight line is

$y = 2x + 3$

Determine (a) the intercept of the line on the $y$ axis, (b) the intercept of the line on the $x$ axis, (c) the gradient of the line and (d) the angle the line makes with the $x$ axis. *Hint:* When the line intercepts the $x$ axis the $y$ coordinate is zero.

**A.5**   Solve the simultaneous equations for the unknowns $x$ and $y$

(a)  $\begin{aligned} 2x + 3y &= 3 \\ x + y &= 1 \end{aligned}$   (b)  $\begin{aligned} 2x + 3y &= 7 \\ 4x - 5y &= 3 \end{aligned}$

**A.6**   Complete Table PA.1.

*Table PA.1*

| θ (°) | θ (rad) |
|-------|---------|
| 30    |         |
| 45    |         |
|       | 1.0     |
|       | 1.5     |
| 135   |         |
|       | 2.2     |

**A.7**   Use Table A.3 to show
(a)  $\sin(\theta - \phi) = \sin(\theta)\cos(\phi) - \sin(\phi)\cos(\theta)$,
(b)  $\cos(\theta - \phi) = \cos(\theta)\cos(\phi) + \sin(\theta)\sin(\phi)$

**A.8**   Making use of the definitions of $\sin(\theta)$ and $\cos(\theta)$ in terms of the sides of a right-angled triangle of sides $a$, $b$ and $c$, deduce Pythagoras' theorem:

$a^2 + b^2 = c^2$

# *Solutions to problems*

## Chapter 1

**1.1** (a) $kg\,m\,s^{-2}$, (b) $s^{-1}$.

**1.2** $Q \times V$ has the units of electric charge multiplied by the units of electric potential, i.e.

$$(A\,s) \times (kg\,m^2\,s^{-3}\,A^{-1}) = kg\,m^2\,s^{-2}$$

Comparison with Table 1.3 shows that this has the same units as energy.

**1.3** (a) $5.932 \times 10^2$, (b) $4.287 \times 10^{-3}$, (c) $2.15 \times 10^{-1}\,A$, (d) $8.5 \times 10^{-5}\,A$, (e) $5.0\,V$, (f) $2 \times 10^{-3}\,F$.

**1.4** (a) $2.5\,\Omega$, (b) $7.5\,M\Omega$, (c) $75\,\mu J$, (d) $0.75\,MV$, (e) $47\,pF$.

**1.5** In standard form: (a) $1.352 \times 10^{-3}$, (b) $7.234$, (c) $8.15 \times 10^3$, (d) $1.71 \times 10^4$.

**1.6** Use $1\,mm = 10^{-3}\,m$, so $1\,mm^3 = 10^{-9}\,m^3$.

## Chapter 2

**2.1** Table P2.1:

*Table P2.1*

| $Q_1$ | $Q_2$ | Sign of force | Repulsive or attractive? | Charge movement? |
|---|---|---|---|---|
| positive | positive | positive | repulsive | move apart |
| positive | negative | negative | attractive | move together |
| negative | positive | negative | attractive | move together |
| negative | negative | positive | repulsive | move apart |

**2.2** (a) $2 \times 10^7\,N$, (b) $-2.5 \times 10^7\,N$, (c) $-7.2 \times 10^5\,N$, (d) $2.8125 \times 10^3\,N$.

**2.3** For $r = 20$ mm, $F_E = 225$ N. Call this force $F_E(20 \text{ mm})$, which just means the force between the charges when the separation is 20 mm. Similarly, call the force between the charges when the separation is 30 mm $F_E(30 \text{ mm})$. Since the charges $Q_1$ and $Q_2$ are fixed, then $F_E \propto 1/r^2$. Let us look at the ratio of these two forces

$$\frac{F_E(30 \text{ mm})}{F_E(20 \text{ mm})} = \frac{1/(30)^2}{1/(20)^2} = \frac{1.111 \times 10^{-3}}{2.5 \times 10^{-3}} = 0.444$$

where we have used the fact that the charges have not changed in value. Therefore, the force between the charges when the distance between them has changed to 30 mm is

$$F_E(30 \text{ mm}) = 0.444 \times F_E(20 \text{ mm}) = 100 \text{ N}$$

When the force acting between the charges has been determined the reader should find the resulting curve has a similar shape to that given in Figure 2.3.

**2.4**  $1.8 \times 10^6$ N.

**2.5**  (a) $1.030 \times 10^7$ N,  (b) $1.315 \times 10^7$ N,  (c) $1.419 \times 10^7$ N,  (d) $1.440 \times 10^7$ N, (e) $1.440 \times 10^7$ N.

**2.6**  Remove $1.875 \times 10^7$ electrons.

**2.7**  1.095 m.


# Chapter 3

**3.1**  (i) $-3.2 \times 10^{-5}$ A $(= -32 \text{ μA})$,  (ii) $-1250 \text{ m s}^{-1}$.

**3.2**  $1.5625 \times 10^{20}$.

**3.3**  In terms of units we find the right-hand side has units

$$\frac{1}{m^3} \times C \times \frac{m}{s} \times m^2 = \frac{1}{\cancel{m}^3} \times C \times \frac{\cancel{m}}{s} \times \cancel{m}^2 = \frac{C}{s}$$

According to Table 1.3 we have 1 C is equal to 1 A s, so that

$$\frac{C}{s} = \frac{A\cancel{s}}{\cancel{s}} = A$$

**3.4**  0.05 C $(= 50 \text{ mC})$.

**3.5**  0.05 A $(= 50 \text{ mA})$.

**3.6**  2 C, equivalent to the flow of $1.25 \times 10^{19}$ electrons.


# Chapter 4

**4.1**  5219 V $(= 5.219 \text{ kV})$.

**4.2**  (a) $1.6 \times 10^{-16}$ J,  (b) $4.38 \times 10^5 \text{ m s}^{-1}$,  (c) $1.6 \times 10^{-16}$ J,  (d) $6.19 \times 10^5 \text{ m s}^{-1}$.

**4.3** $8.35 \times 10^{-14}$ J, 521.9 kV.

**4.4** (a) 2 J, (b) work done to move charge across $R_1$ 1.6 J, across $R_2$ 0.4 J, across $R_3$ 0.4 J.

**4.5** From Table 1.1 $mv^2/2$ has the SI unit (kg) $\times$ (m s$^{-1}$)$^2$ = kg m$^2$ s$^{-2}$. Similarly, from Table 1.3 $Q \times V$ has the SI unit (A s) $\times$ (kg m$^2$s$^{-3}$A$^{-1}$) = kg m$^2$s$^{-2}$. From Table 1.3 both of these quantities have the SI unit of energy.

# Chapter 5

**5.1** 0.25 $\Omega$.

**5.2** 0.547 $\Omega$, 0.137 $\Omega$.

**5.3** 6.06 A.

**5.4** $8.6 \times 10^{-7}$ m$^2$ = 0.86 mm$^2$.

**5.5** $\ell_{Cu}$ = 29.1 m, $\ell_{Ni}$ = 0.45 m.

**5.6** 65 cm$^2$.

**5.7** $2.5 \times 10^{-3}$ $\Omega$ m.

**5.8** $1.885 \times 10^{-3}$ $\Omega$ m.

**5.9** 100$\rho$ $\Omega$.

**5.10** $1.57 \times 10^{-8}$ $\Omega$ m, $1.65 \times 10^{-8}$ $\Omega$ m, $1.94 \times 10^{-8}$ $\Omega$ m.

**5.11** 121°C, $-78$°C, $-105$°C.

**5.12** 0.05 (°C)$^{-1}$, 40°C.

**5.13** Table P5.1.

*Table P5.1*

| 1st band | 2nd band | 3rd band | 4th band | Nominal resistance | Tolerance | Resistance range |
|----------|----------|----------|----------|--------------------|-----------|-------------------|
| brown | black | brown | gold | 100 $\Omega$ | 5% | 95 $\Omega$,105 $\Omega$ |
| red | red | green | silver | 2.2 M$\Omega$ | 10% | 1.98 M$\Omega$, 2.42 M$\Omega$ |
| green | blue | green | gold | 5.6 M$\Omega$ | 5% | 5.04 M$\Omega$, 6.16 M$\Omega$ |
| red | violet | orange | silver | 27 k$\Omega$ | 10% | 24.3 k$\Omega$, 29.7 k$\Omega$ |
| brown | black | brown | silver | 100 $\Omega$ | 10% | 90 $\Omega$, 110 $\Omega$ |
| green | blue | green | none | 5.6 M$\Omega$ | 20% | 4.48 M$\Omega$, 6.72 M$\Omega$ |

**5.14** (a) brown, green, brown, silver, (b) red, red, orange, gold, (c) red, yellow, yellow, (d) brown, black, black, silver.

**5.15** Nominal current = 0.1 A, maximum current = 0.111 A, minimum current = 0.091 A.

# Chapter 6

**6.1** 4 $\Omega$ and 6 $\Omega$.

**6.2** (a) 10 Ω,  (b) 33.6 Ω,  (c) 38.6 Ω.

**6.3** (a) 1/3 A,  (b) 1/4 A,  (c) 1/4 A,  (d) 1/12 A,  (e) the p.d.s follow from using $V = IR$.

**6.4** $I_1 = 2.5$ mA, $I_2 = 7.5$ mA, $I_3 = 10$ mA.

**6.5** Without ammeter $I = 0.714$ A, with ammeter $I = 0.625$ A.

**6.6** (a) 0.5 A,  (b) 10 V,  (c) 9.091 V,  (d) 9.980 V.

**6.7** 4 Ω.

**6.8** 32 V.

**6.9** $R = 8.667$ kΩ, $I = 1.154$ mA.

**6.10** $E = 1.5$ V, $r = 0.5$ Ω.

**6.11** $r = 1$ Ω.

**6.12** Maximum power dissipated by $R$ when $R = 20$ Ω.

# Chapter 7

**7.1** $V_2 = 8$ V, $V_3 = 2$ V, $V_5 = 0$ V. The current flowing through each resistor is given by Ohm's law.

**7.2** $I = -1$ A.

**7.3** $I(20\,Ω) = 5/11$ A, $I(30\,Ω) = 1/11$ A, $I(10\,Ω) = 4/11$ A, although care has to be taken with regard to the sign convention chosen for current flow.

**7.4** $I(12\,Ω) = 14/3$ A, $I(3\,ΩQ) = 0$ A, $I(6\,Ω) = 14/3$ A, although care has to be taken with regard to the sign convention chosen for current flow.

**7.5** $V_{AB} = 20/3$ V.

**7.6** $E = 12.5$ V.

**7.7** Should find $V_{Th} = 8.333$ V, $R_{Th} = 16.667$ Ω (be careful with the sign on $V_{Th}$).

**7.8** $R_{Th} = 8.571$ Ω.

**7.9** Should find $V_{Th} = 39.2$ V, $R_{Th} = 2.4$ Ω (be careful with the sign on $V_{Th}$).

# Chapter 8

**8.1** $B(X) = 5 \times 10^{-6}$ T, $B(Y) = 2.5 \times 10^{-6}$ T, $B(Z) = 1.667 \times 10^{-6}$ T. The direction of the magnetic field at these points is given by the right-hand grip rule.

**8.2** (a) $F = 0.016$ N,  (b) $F = 0$.

**8.3** (a) $F_1 = 3.333 \times 10^{-4}$ N, $F_2 = 6.0 \times 10^{-4}$ N, $F_3 = 2.667 \times 10^{-4}$ N.

**8.4** 3.77 N.

**8.5** 50 μN m.

**8.6** $v = 1.875 \times 10^7$ m s$^{-1}$, $r = 0.053$ mm, $F = 6 \times 10^{-12}$ N.

**8.7** $1.761 \times 10^{11}$ C kg$^{-1}$.

## Chapter 9

**9.1** (a) $\mathscr{E} = 3$ V.

**9.2** (a) 0.491 Wb, (b) 0.982 Wb, (c) 196.350 V, (d) 196.350 V.

**9.3** 0.509 T.

**9.4** $\mathscr{E}_{AB} = 0$ V, $\mathscr{E}_{CD} = 3.75$ V, $\mathscr{E}_{EF} = 0$ V.

**9.5** Maximum value = 9.42 V, minimum value = $-9.42$ V.

## Chapter 10

**10.1** 0.5 H.

**10.2** $L = 100$ H, $\varepsilon = 2\,000$ V.

**10.3** $L = 1.579 \times 10^{-6}$ H.

**10.4** According to Table 1.3, $L/R$ has the units of

$$\frac{\text{kg m}^2\,\text{s}^{-2}\,\text{A}^{-2}}{\text{kg m}^2\,\text{s}^{-3}\,\text{A}^{-2}} = \frac{\text{s}^{-2}}{\text{s}^{-3}} = \text{s}$$

**10.5** (a) 1 H, (b) 0.5 mA, (c) 0.1 ms, (d) 0.0693 ms, (e) 0.031 250 μJ,
(f) 625 μW, (g) Table P10.1:

*Table P10.1*

| $t$ (ms) | $I$ (mA) | $V_R$ (V) | $P$ (mW) | $V_L$ (V) | $dI/dt$ (mA/s) |
|---|---|---|---|---|---|
| 0 | 0.0 | 0.0 | 0.0 | 5 | 5 000 |
| 0.05 | 0.197 | 1.967 | 0.387 | 3.033 | 3 033 |
| 0.15 | 0.388 | 3.884 | 1.509 | 1.116 | 1 116 |
| 0.25 | 0.459 | 4.590 | 2.106 | 0.410 | 410 |
| 0.50 | 0.497 | 4.966 | 2.466 | 0.034 | 34 |
| 1.00 | 0.500 | 5.000 | 2.500 | 0.000 | 0 |

**10.6** Step-up transformer, 0 V, 625 V.

**10.7** 66.667 V.

## Chapter 11

**11.1** 10 ms.

**11.2** 12.5 mA.

**11.3** 50 nF.

**11.4** $C = 885.4$ pF, (a) 8.854 nC, (b) 200 V, (c) initial and final energies are 44.3 nJ and 885 nJ, respectively.

**11.5** 25 μF.

**11.6** $C = 20.833$ μF, $Q$ on $C_1 = 0.25$ mC, $Q$ on $C_2 = 1.25$ mC, p.d. across each capacitor = 10 V.

**11.7** 2.656 nF.

**11.8** $C = 5$ μF, $Q = 100$ μC, $V_{(10\,μF)} = 10$ V, $V_{(5\,μF)} = 20$ V.

**11.9** $I_{AB} = 0$, $I_{CD} = 0.750$ mA, $I_{EF} = 0$.

**11.10** According to Table 1.3, $RC$ has the units of

$$(\text{kg}\,\text{m}^2\text{s}^{-3}\text{A}^{-2}) \times (\text{kg}^{-1}\text{m}^{-2}\text{s}^4\text{A}^2) = \text{s}^{-3} \times \text{s}^4 = \text{s}$$

**11.11** (a) 50 ms, (b) 1.81 V, (c) 8.19 V, (d) 5.27 ms, (e) 7.27 ms.

**11.12** Answers the same as in Q11.11.

# Chapter 12

**12.1** $\mathscr{E}_0$ (1 turn) = 0.094 2 V, $\mathscr{E}_0$ (100 turns) = 9.42 V. This gives the peak value as 9.42 V, the peak-to peak value as 18.84 V and the r.m.s. value as 6.661 V. The frequency and the period are $f = 50$ Hz and $T = 0.02$ s, respectively.

**12.2** (a) $f = 0.2$ Hz, $T = 5$ s, $\langle \mathscr{F} \rangle = 100$, (b) $f = 5$ Hz, $T = 0.2$ s, $\langle \mathscr{F} \rangle = 0$.

**12.4** $\mathscr{E}(t) = A \sin(200\pi + \Phi)$ with $A = 13.229$ V, $\Phi = 0.714$ rad.

**12.5** $X_C = 6\,366$ Ω, peak current = 0.157 A, $X_L = 62.832$ Ω, peak current = 1.592 A.

# Chapter 13

**13.1** $Z = 1250$ Ω, $X_L = 1000$ Ω, $L = 3.183$ H, $I(t) = 20 \sin(100\pi - 0.927)$ mA.

**13.2** $X_C = 63.662$ Ω, $Z = 118.544$ Ω, the current is $I(t) = 0.422 \sin(500\pi t + 0.567)$ A.

**13.3** For $f > f_0 = 712$ Hz we have $X_L > X_C$ so that the phase shift is $\pi/2$, while for $f < f_0 = 712$ Hz we have $X_L < X_C$ so that the phase shift is $-\pi/2$.

**13.4** According to Table 1.3, $LC$ has the units of

$$(\text{kg}\,\text{m}^2\text{s}^{-2}\text{A}^{-2}) \times (\text{kg}^{-1}\text{m}^{-2}\text{s}^4\text{A}^2) = \text{s}^{-2} \times \text{s}^4 = \text{s}^2$$

so that $\sqrt{(LC)}$ has the units of time.

**13.5** 250 Ω.

**13.6** (a) $X_L = 62.832$ Ω, $X_C = 79.577$ Ω, $Z = 26.085$ Ω, (b) $I = I_0 \sin(2000\pi t - \Phi)$, with $I_0 = 0.383$ A, $\Phi = -0.697$ rad (note that the current *leads* the supply voltage since $\Phi$ is negative), (c) $\cos(\Phi) = 0.767$, $P_{ave} = 1.468$ W, (d) $f_0 = 1125$ Hz (= 1.125 kHz).

## Chapter 14

**14.1** (a) $I = 0.47$ A, (b) $I = 0.5$ A, (c) $I = 0$.

**14.2** (a) $I_{(35\,\Omega)} = 0$, $I_{(5\,\Omega)} = 0.5$ A, $I_{(10\,\Omega)} = 0.5$ A, (b) $I_{(35\,\Omega)} = 0.125$ A, $I_{(5\,\Omega)} = 0$, $I_{(10\,\Omega)} = 0.5$ A, (c) $I_{(35\,\Omega)} = 0$, $I_{(5\,\Omega)} = 0.44$, $I_{(10\,\Omega)} = 0.5$ A.

**14.3** Half wave rectifier: $f = 500$ Hz, $\langle V \rangle = 10/\pi$ V $= 3.183$ V. Full wave rectifier: $f = 1000$ Hz, $\langle V \rangle = 20/\pi$ V $= 6.366$ V.

## Chapter 15

**15.1** $I_B = 0\ 047$ mA, $I_C = 7.05$ mA, $I_E = 7.097$ mA, $I_C \times R_C = 7.05$ V, $V_{CE} = 2.95$ V.

**15.2** $I_B = 0.022$ mA, $I_C = 11$ mA, $I_E = 11$ mA, $I_C \times R_C = 11$ V, $V_{CE} = 4$ V.

**15.3** $I_B = 0.108$ mA, $I_C = 10.8$ mA, $I_E = 10.9$ mA, $I_C \times R_C = 10.8$ V, $V_{CE} = -0.8$ V. Since the collector–emitter voltage has turned out negative our initial assumption that the transistor is operating in its active region is incorrect.

## Appendix

**A.1** (a) $x = 2.5$, (b) $x = 200$, (c) $x = 0.8$, (d) $x = -0.375$.

**A.2** (a) $x = 2$, (b) $-0.75$, (c) $x = 1$.

**A.4** (a) 3, (b) $-1.5$, (c) 2, (d) 1.107 rad ($= 63.435°$).

**A.5** (a) $(x, y) = (0, 1)$, (b) $(x, y) = (2, 1)$.

**A.6** Table PA.1:

*Table PA.1*

| θ (°) | θ (rad) |
| --- | --- |
| 30 | 0.524 |
| 45 | 0.785 |
| 57.296 | 1.0 |
| 85.944 | 1.5 |
| 135 | 2.356 |
| 126.051 | 2.2 |

# *Index*